PHOTONIC TECHNOLOGY
AND
INDUSTRIAL POLICY

PHOTONIC TECHNOLOGY
AND
INDUSTRIAL POLICY

U.S. Responses to Technological Change

ERNEST STERNBERG

State University of New York Press

Published by
State University of New York Press, Albany

©1992 State University of New York

All rights reserved

Printed in the United States of America

No part of this book may be used or reproduced
in any manner whatsoever without written permission
except in the case of brief quotations embodied in
critical articles and reviews.

For information, address the State University of New York Press,
State University Plaza, Albany, NY 12246

Production by Christine M. Lynch
Marketing by Bernadette LaManna

Library of Congress Cataloging-in-Publication Data

Sternberg, Ernest.
   Photonic technology and industrial policy : U.S. responses to technological change / Ernest Sternberg.
      p.   cm.
   Includes bibliographical references (p. 291) and index.
   ISBN 0-7914-1181-8 (ch : acid-free). — ISBN 0-7914-1182-6 (pb : acid-free)
   1. Optoelectronics industry—Government policy—United States.
   2. Laser industry—Government policy—United States.   3. Fiber optics industry—Government policy—United States.   4. Image processing equipment industry—Government policy—United States.
   5. Computer industry—Government policy—United States.
   6. Photonics—Industrial applications—Research-Government policy—United States.   7. Photonics—Research-Government policy—United States.   I. Title.
   HD9696.O673U675   1992
   338.4'762136—dc20   91-34657
                                                              CIP

10 9 8 7 6 5 4 3 2 1

For my parents

CONTENTS

Acknowledgments / ix

1	A Technology and an Argument Introduced	1
2	The Rise of Photonic Technology	15
3	Technology as a Common Resource	45
4	Privatization and the Industrial Policy Debate	73
5	Disaggregation	113
6	Collaboration	141
7	Sheltering	175
8	Conclusions and Implications	227

Notes / 239

Select Bibliography / 291

Index / 299

TABLES

2-1 Japanese Production in Optoelectronics 33
2-2 Sectors to be Affected by Optoelectronic Technology 34
2-3 U.S. Patents Granted in Lightwave Technology 38
2-4 Estimates for Optoelectronic R & D Expenditures 40
6-1 University Research Centers in the United States in Photonics 145
7-1 U.S. Department of Defense Science and Technology Funding for Twenty-two Critical Technologies 186

ACKNOWLEDGMENTS

I wish to thank my friends, colleagues, and teachers whose advice, suggestions, and criticisms have made this book possible.

It was at the Nelson A. Rockefeller Institute of Government in Albany, New York, that I first considered studying industrial policy and advanced optics. At this fine institution, I was given the time and freedom to reflect on public policy issues and conduct the readings that form the theoretical core of this study.

Thereafter I was fortunate to be able to work at the Center for Governmental Research in Rochester, New York. There I had the opportunity to work with a community of persons interested in the future of optics and imaging technology. I am grateful to the Center for giving me the leeway to pursue my own studies while also conducting research projects more specifically related to Rochester's economic development. I would particularly like to acknowledge the many fruitful conversations with Alan Taddiken about photonics and U.S. industrial policy.

In the course of conducting my research, I interviewed numerous persons working in university research centers, government agencies, and a few private firms. They shared with me descriptions of their work or their organization, helping give substance to my ideas. Where I directly quote or use bits of factual information, I cite the interview in the text. While my interlocutors knew of the general topic of my research, they were unaware of the arguments I was constructing. Though it is clear, it is worth stating that neither they nor any of the persons who commented on the manuscript necessarily share my opinions or agree with my interpretations.

I am indebted to several colleagues who gave valuable comments on early drafts of the study. They include Davydd J. Greenwood, Michael Heiman, Judith Reppy, Michael Schwartz, James White, and Randall P. Wilson. Brian J. Thompson of the University of Rochester kindly took time from his busy schedule to make corrections and recommendations. He and I remain, however, in friendly disagreement over some points of detail and interpretation. Thanks go also to an anonymous reviewer for State University of New York Press, whose constructive suggestions helped improve the manuscript.

And I am grateful to Mitchel Y. Abolafia, John Forester, and Theodore J. Lowi, my advisers at Cornell University. Their encouragement and acute comments led me from initial proposals through writing and rewriting. Theodore Lowi was a steadfast guide through the politics of public policy, the strains of manuscript revision, and the vagaries of publishing. His help and continuing encouragement are very much appreciated. And I am very fortunate to have worked with John Forester. My conversations with him indeed helped shape this book, but they also left me with an enduring insight into critical thinking. His patient questioning and his view of this work as a sustained argument guided me through intellectual labyrinths. He taught me how to become a scholar.

As I close the work, my love goes to Zohi, to Elie and Danny, and to my parents, to whom it is dedicated.

ONE

A Technology and an Argument Introduced

PHOTONICS AND ITS INDUSTRIAL REVERBERATIONS

In the general material clutter of our daily lives, photonics has just begun to make an inroad. The compact disk player, in which a laser device replaces a phonographic needle, has become a ubiquitous middle class possession. Fiber-optic lines stretch across the continent and the oceans. Optical computer memories encode small reference libraries on plastic disks weighing an ounce or two. Cash register clerks fumble with bar codes to be read by laser scanners. Hospital patients are subjected to laser surgery and fiber-optic catheterization, among other novelties.

Beyond the horizons of daily life, satellites relay digital images of weather patterns and missile emplacements. Machine-vision systems work in factories as diligent if unversatile product inspectors. Using digital cameras, which record images on a small diskette, journalists can electronically convey a photograph to an editor. Pilots of stealth bombers, flying at night, view distant battlefields and invisible enemies as hologram images projected on the windshield.

Technological prognosticators tell of integrated optoelectronic systems supplying homes and businesses with services that combine communications, computing, entertainment, high-definition images, and information retrieval. Not much further in the future, the most prodigious of computing machines, the optical computer, might also become possible, tying together fiber-optic communications and information processing services at the speed of light.

What these new technical possibilities have in common is that they work through the application of light. The prospect of new industries founded on optics has caused considerable excitement. None the wearier for the space age and atomic age, *The Futurist* and *Time* magazines have already heralded the age of light.[1] But even those who would hesitate at welcoming still another technological age might acknowledge that the new technology, *photonics*, is setting off reverberations throughout the world economy.

Technological change of such magnitude has effects running the breadth of human affairs, effects medical and environmental, military and industrial. The technology yields imaging devices of fine resolution for medical diagnosis and equipment for less invasive surgical procedures. At the same time, in the search for materials that simultaneously carry electricity and light, the manufacture of exotic new optoelectronic materials adds to the world's stock of environmentally hazardous substances. The technology also has momentous military significance. A variety of night-vision, range-finding, guidance, detection, and display devices depends on photonics. Optoelectronic imaging is essential to the satellite reconnaissance systems that make arms control verification possible, and also to the guidance systems of low-flying intermediate range missiles that can destabilize arms control. These very techniques yield enormous quanitites of data, which themselves create demands for the rapidity of optical communications and, if they can be perfected, optical computers.

Most of all, photonics has consequences for numerous industries. They include core industries making specialized optoelectronic components, lasers, image-capture devices, and optical components. They also include industries making final products: computer peripherals, medical equipment, telecommunications equipment, avionic instrumentation, automatic inspection systems, home entertainment appliances, scanning devices, and document imaging systems. And these devices and systems often prove to be critical elements in the productivity of still other industries: retailing, insurance, manufacturing processes, telecommunications, and office work of every description.

The industrial blessings are mixed, as usual. The technology creates new industries making optical cable, optical computer memories, and optoelectronic imaging systems, but it cuts into older industries such as copper cable and satellite communications, magnetic computer memories, and chemical photography.

Since the technology transforms production in almost every branch of the economy, photonics already qualifies to be classified with the select company of technological revolutions that have transformed twentieth-century industrial society. Photonics takes its place alongside internal-combustion technology, pharmaceuticals, electricity, petrochemicals, and microelectronics. It also belongs alongside the more revolutionary of the current generation of destabilizing technologies, such as advanced ceramics, superconductivity, and biotechnology. Photonics is, however, more advanced in its industrial effects than are the other recent technological entrants. It has already attained worldwide commercial significance.

Since 1980, the multifarious technical possibilities along with military, industrial, and consumer uses for the new products have drawn numerous firms to photonic technology. By 1988, a directory listed more than twenty-two hundred domestic companies and divisions, small and large, with an interest in photonics. These American firms, along with foreign ones, represented considerable industrial production. A rough estimate from various sources suggested that by 1989 the worldwide market for products predominantly dependent on photonic technology was in the $15–20 billion range. It was a much larger commercial market than that of another and better known technological development, biotechnology.[2]

According to Japanese figures, that country's production of optoelectronic goods increased from $350 million in 1980 to more than $8 billion in 1987. The increase seems greater in dollars than in yen, since the yen's value in U.S. dollars rose sharply in 1986. Nevertheless, even in terms of yen, production in Japan had an average annual growth rate of 50 percent through 1986. Japan's optoelectronic trade association has asserted that this was a rate of growth greater than that of any other Japanese industry.

It will come as no surprise that America's contemporary economic nemesis forged ahead in photonics. Optoelectronics figured prominently in Japan's industrial growth in the 1980s, and Japan became the primary supplier of the world's optoelectronics. By contrast, the United States showed a middling industrial performance. It did well in some years in the export of optical fiber, but progressively lost its relative share of world sales in other photonic products.

By the late 1980s, several sets of data revealed that photonics had become another of the technological sectors in which the U.S. was showing declining industrial competence. Foreign shares of U.S. patents in light-wave communications technology grew from 21 percent in 1973 to more than 50 percent in 1987. The Japanese increased their production of technical papers in the field much faster than U.S. scientists did. Panels of American experts comparing Japanese and U.S. developments in the field reported Japanese advances that were significantly ahead of those in this country. A National Research Council report on photonics concluded that the United States was "already a follower—or, worse, an observer—in developing many of the commercial products of the field."[3]

Since skills in photonics are essential to the industries taking advantage of the new field, it would seem reasonable to suppose that investment in research and development and a history of technical strength in a field would have much to do with a nation's success in the technology. Yet, though lasers, fiber optics, and optoelectronic imaging had largely domestic origins, U.S. industrial capability in the field was being overtaken by that of Japan and Western Europe. Japan did better with photonics despite a research investment that was, at least initially, smaller than that of the United States.

In the critical early years of the 1980s, when industrial possibilities were still open, Japanese industrial research expenditure on optoelectronics only moderately exceeded that in the U.S. When U.S. industrial, military, and university research expenditure is added together (though with rough estimates from disparate sources), domestic research investment turns out to have been much greater than that in Japan. Faltering U.S. industrial strength in photonics was, then, not primarily attributable to the lack of domestic research investment.

The relative decline of U.S. competence in photonics despite substantial domestic expenditure leads to our central problem: Why does a capitalist nation, presumably dependent to a large extent on the market for technological choices, perform worse than others in gaining industrial advantage from a revolutionary new technology?

THE ARGUMENT, VERY BRIEFLY

The chapters that follow contend that we can answer this question after we come to understand photonics as a technological paradigm. In contrast to the widely held conception of technology as an assemblage of discrete artifacts and inventions, the idea of paradigms has us think of it as having integral properties.

As it came into widespread use in the 1980s, photonics was becoming an integral body of technical skill and knowledge. In optical communications and in imaging applications, it was being implemented as integrated systems of devices and software. And it was bringing about a set of technological relationships among diverse industrial sectors. Photonics came to have integral characteristics in all these respects—as a body of technical knowledge, as technical systems, and as technological interrelationships among industries. Photonics acquired its massive importance in the world economy in the 1980s by dint of its properties as a technological paradigm.

If a technology is seen from the usual perspective, as discrete artifacts and pieces of intellectual property, we can reasonably conceive of these items being efficiently traded on markets. But if industrial change is better seen as the outcome of the emergence of a technological paradigm, then the argument for an unfettered market response becomes problematic. Private firms operating in markets might in themselves respond inefficiently to the technological interdependencies inherent to paradigms. In an internationalized economy, firms would operate more successfully in the presence of industrial policies that recognize the integral properties of technology and plan for massive technological changes in the economy.

The idea of technological paradigms, then, gives us a rudimentary criterion by which to assess public policies toward technological sectors: those policies are good that can recognize integral technological relationships in the economy. If so, we can plausibly explain lagging industrial performance in photonics by investigating the policies through which the U.S. has responded to the technology.

We might first expect that the U.S. responded through market solutions. Indeed, on grounds of the allocative superiority of free markets in responding to technological change, an important domestic debate in the 1980s about declining technological competitiveness led to the rejection of industrial policy. But in actuality the U.S. responded only in part through market solutions to technological change. Recognizing that U.S. retrogression in the technology occurred to mutual detriment, corporate, academic, and military leaders sought a more concerted response. Amid widespread rejection of industrial policy as faulty economics, and in the absence of an intellectual means of understanding the integrality of technological change, their response to photonics occurred not through unfettered markets, nor through explicit public policy, but through a *privatization* of policy-making.

Privatized policy took on three forms. First, government assets were disaggregated unit by unit to interested parties through pork-barrel appropriations, business participation in agency operations, and review committees representing eventual beneficiaries. Second, collaborative committees of university faculty and corporate affiliates used public funds to set the technological research agenda, but without accountability to public concerns about technological change and its industrial implications. And third, sector-specific technological research programs, chosen in part with civilian industrial effects in mind, took place in the political shelter of the military establishment.

Privatized industrial policy did not surrender policy-making in favor of market solutions but rather shifted an interventionist economic policy into the shadowy realms of pork-barrel politics, public-private partnerships, and military bureaucracies. This privatization made for debilitating policy. Operating in the absence of vision or strategy, it failed to respond to the integral characteristics of photonics understood as a technological paradigm.

U.S. industrial retrogression in photonics, therefore, reflects the domestic policy-making inability to respond coherently to technological change. To the extent that the U.S. has responded similarly to other technological paradigms and to other economic resources (such as skill and physical infrastructure) having integral qualities, we can draw from the case of photonics the broader lesson that the decline of American industrial capitalism in the 1980s occurred because of the domestic inability to plan.

SOURCES AND LIMITATIONS

The argument so briefly summarized above rests on a case study of the public response to the rise of photonic technology in the United States, especially in the years 1980–90.

The start of the decade saw the introduction of the word *photonics* in trade publications and research laboratories to express the technological possibilities seen in the convergence of traditional optics, lasers, fiber optics, optoelectronic imaging, and optoelectronic computing. By the end of the decade, the technology was absorbed into the operations and strategies of numerous industries and accounted for a substantial worldwide industrial sector. Though the technology had origins going back to the previous century, its full industrial flowering occurred in the decade of the 1980s. Importantly for the present argument, it was the same decade that saw widespread recognition of the internationalization of the U.S. economy and growing concern over declining U.S. industrial competence in world markets.

In looking at the public response to photonics in this decade, this study by and large restricts itself to research and development policy. This is not a desirable limitation, since the spread of the industrial applications of photonics is shaped by several kinds of policies. These include policies addressing technical training and advanced engineering education, military acquisitions, the deregulation of telecommunications, technological export controls, and technical standardization, as well as R & D. But if the study was to be practicable, its scope had to be narrowed. Hence, the limitation to R & D. Even the discussion of the U.S. agency that deals with standardization will concentrate on the agency's R & D effort, and not on standardization policy, though that in itself is an important component of industrial policy.

Unlike biotechnology and semiconductor technology, photonics (by that name or under other rubrics, such as optics) has remained uninvestigated in social-science literature. The present study therefore had to find its material in original sources. And since the subject is not well recognized and cataloged with regard to questions of public policy, the sources are various and disparate. Prominent among them are interviews with governmental, industrial, and university professionals concerned about photonics. Other sources include congressional hearings, articles in the general press, and back issues of the trade press (*Laser Focus World, Lasers and Optronics, OE Reports, Optics and Photonics News,* and *Photonics Spectra* by these or earlier names). Still other sources included a variety of what librarians call "fugitive materials": U.S. and foreign government documents, consultants' reports, brochures, draft documents, and so forth.

Now and then in the text, references appear to Rochester, New York, and its academic institutions. In part, Rochester's prominence just reflects that the research on this study was conducted there. More importantly, Rochester belongs in the text, since it continues as the traditional home of American optics, especially in its applications to imaging.

The argument that follows emerges, therefore, from an explicitly restricted scope of investigation. The scope encompasses the responses of one nation, the United States, to the emergence of one resource, photonic technology, in one decade, from the vantage of an author living in one city. The observations to be made all suffer from the limitations of a case study. The limitations being understood, the study builds on this case a broader argument on U.S. public response to technologies (and other resources, such as education and physical infrastructure) that are the common resources of capitalist industrial production.

THE ARGUMENT IN OUTLINE

The book is divided into eight chapters, of which the first is the present introduction. Chapter 2 defines photonics, traces its history, shows its commercial importance, and gives evidence of lagging U.S. performance in the technology relative to other nations, particularly Japan.

Chapter 3 argues that photonics should be considered a technological paradigm, because it exhibits integral properties of three kinds. First, photonics is (or is becoming) an integral body of knowledge, technical skill, and engineering practice. Second, photonics exhibits the properties of a system, because it comes into practical use through networks of interrelated devices and software. And third, the technology interweaves numerous industries that have become dependent on it or vulnerable to it. The photonics paradigm, therefore, cannot be properly understood as an assemblage of individual innovations.

Technological paradigms are not unique in having such properties. Paradigms fall under a broader class of common resources that includes physical infrastructure and educational and training resources. In these other collective resources, as in technological paradigms, the development of the resource becomes knowable. Nations (and regions) can use such knowledge to formulate public policy that responds to the development of the resource—to technological development in our case—and its industrial implications.

In a capitalist state equipped with institutions that can use such planning knowledge for coherent allocation of technological resources, private firms are better prepared than are firms in other states to take industrial advantage of a new technology. The concept of common resources then provides us with the principle by which we can assess policy responses to photonic technology.

As chapter 4 relates, photonics achieved its industrial importance in the 1980s, a decade of U.S. industrial retrogression relative to other capitalist nations. Since a nation accustomed to high standards of living would especially need to maintain its strength in technology-intensive production, the decade's declines in technologically advanced industries have been seen as the most disturbing. American industrial resurgence might then hinge on technology. What role, if any, should public policy have in reinvigorating technology?

The question occasioned a short-lived debate about industrial policy as a potential response to the decline. It was the outcome of the debate that while industrial policy was rejected by federal policymakers and mainstream intellectuals, it was widely implemented, but in sublimated, privatized form. It came to be known as "industry-university cooperation," "industry-led policy," "competitiveness policy," "defense industrial base policy," and "defense technology policy"—the euphemisms of privatization.

This was surely a curious result. By cherished belief, the U.S. would be expected to take the path mandated by tradition and by widespread depictions of its political economy. The U.S. would, in this view, act consistently with liberal economic principles and seek market solutions to technological change. If there were also to be ad hoc subsidies or protectionist measures, these were more or less frequent, more or less harmful exceptions. The political triumph of the opponents of industrial policy would be expected. It fit a capitalist state known for its market ideology.

To the contrary, however, the privatized response to photonics did not mean a return of technological decisions to the market. It meant, instead, that an interventionist economic policy came to be pursued under public-private, private, and military auspices.

To critics of industrial policy, such an outcome is distressing mainly because such policy is, from the start, misconceived economics. The privatization of policy-making then represents a species of bungling, just the sort of effect to be expected when government is given industrial responsibilities. Such policy is misguided less by dint of privatization than by the wrongheadedness of the policy itself.

If our assessment is to be any different from that of the opponents of industrial policy, then we must confront and find a flaw in their resolution of the industrial policy debate. Such an encounter should start with a definition. Industrial policy may be best defined as policy toward specific industrial sectors that operates through knowledge of, and judgment about, those sectors. The definition highlights the very issue on which the debate turned: the ability of government to make decisions that are better than those of the committed entrepreneur, technical researcher, and investor.

Advocates of industrial policy contended that certain sectors, especially certain critical technologies, were strategic to the broader performance of the economy, so they required special attention in policy-making. The opponents of industrial policy retorted that government policy might then have to target some technologies or industries, a process that would require detailed knowledge of economic sectors. Government bureaucrats, as their counter-argument went, would be put in a position of having to intelligently choose among sectors or even engage in planning, though such intelligence had frequently escaped them on previous occasions.

The opponents of industrial policy, therefore, posed as their paramount objection an argument about knowledge. They held that government would have less knowledge than private actors have in responding to technological opportunities. Government policymakers would then surely provide an inferior substitute for the technological decisions of self-motivated researchers, research managers, and venture capitalists.

But if superior private knowledge was the grain of the argument, an ideological presupposition was its kernel. The idea that private actors have better knowledge than public ones in technological choices makes the unexamined presumption that technology consists of a set of discrete inventions. When technical progress is conceived of as discrete inventions, private firms and inventors are no doubt the ones best suited to deciding which research projects and incipient inventions deserve further effort and have commercial potential. The argument about knowledge, as used by the opponents of industrial policy, rises or falls on the truth of this assumption.

The assumption is breached, however, if technological transformations like photonics have integral properties that exceed the intentions of individual decision-makers. In its integrality, technological change gains a measure of predictability. Individual decision-makers may become better at making technical decisions when they operate in a nation where technology is recognized as having such integral properties and where public policies anticipate and respond to technological change. If so, then we could assess privatized industrial policy by its ability to learn about, respond to, and plan for the structured relationships between technologies and industry.

Hence, chapter 4 provides us with a principle by which we can assess privatized industrial policy. By this principle, we should first ask whether an economic policy is indeed industrial policy (whether it operates through sector-specific knowledge). Then we should ask how privatization worked—how this unexpected kind of policy-making was actually carried out. Finally, we should ask whether such privatized sectoral policy had the ability to undertake the very kind of task that would give it an economic justification within capitalism, the task of responding with planning knowledge to the emergence of a technological paradigm as a common resource.

These questions pose a challenge to the rest of the study. The next three chapters, chapters 5 through 7, set out to assess privatized policy-making. At the risk of excessive schematization, such policy-making is divided into three forms: disaggregation, collaboration, and sheltering.

In *disaggregation*, public funds for technological research in photonics are distributed among beneficiaries, such as university researchers or research-performing businesses, that hold the most direct interests in research. In *collaboration*, a newer class of policies emerging largely since the late 1970s, centers of technical strength are established for the allocation of research resources. In these centers, the research agenda is set by industry-university or all-industry groupings that operate without formal governmental accountability for their technology policy decisions. Recognizing that there are overlaps in the technological needs of industry and the military agencies and that the military agencies afford some protection from factious politics, policy-makers and military bureaucrats also practice *sheltering*. In this form of industrial policy, research endeavors, meant to a significant extent for industrial purposes, are carried out under military agencies.

Disaggregation, collaboration, and sheltering, then, represent the three forms of privatization. But they do not constitute hard and fast distinctions. Every intermediate gradation and every combination of these forms has also been tried.

Discussion of the three forms of privatization is arranged roughly in ascending order, from a lesser to somewhat greater level of organizational capability to respond coherently to the rise of photonics. In disaggregation, examined in chapter 5, government's assets are distributed in cash or kind to universities, firms, and academic researchers through disparate decisions that bear no relationships to each other or to any articulated conception of what is to be accomplished with photonics. Three kinds of disaggregation are discussed: pork-barrel appropriations, business participation in government labs, and review panels that examine funding proposals according to criteria that assess technical merit.

In each kind of disaggregative policy-making examined, someone, somewhere made decisions about the dispersal of public funds specifically for photonics. Means included the strategic lobbying decisions of universities and their consultants, the log-rolling of the budgetary process, mutual consultations between agency employees and business participants, and the research funding decisions of "merit review" panels composed of industrial and academic participants.

Each form of decision-making had international competitiveness in photonics as an expressed purpose. But research policy decisions toward this end were made in large part by the potential beneficiaries of the policy, each beneficiary operating more or less in isolation from the others. None of the decision-makers sought or were given access to a broad conception of the industrial directions that photonic technology would take, except for such ideas that the participants may have collected through serendipitous experience. Disaggregative policy-making was indeed a form of privatized industrial policy directed at photonics, but it operated without any general knowledge of purposes or directions.

In the collaborative form of privatization, discussed in chapter 6, industrial consortia and industry-university committees made R & D decisions. Industrial research consortia came to have such authority in the 1980s when new federal legislation freed them to pursue joint research. Industry-university research centers were a more important kind of U.S. response to photonics. In the 1980s, federal and state governments established numerous university research centers specifically for photonics (or optics, optoelectronics, or fiber optics) research for the sake of industrial competitiveness.

Inevitably, committees composed of industrial affiliates and academic researchers served on boards of directors and as policymakers for the centers. These committees allocated research funds in the absence of a public role in the allocation decision.

The industry-university joint research centers differed from industrial participation in government labs and from anonymous peer review in that the centers potentially created productive interrelationships among researchers and could allocate funds according to some vision of industrial purposes. The knowledge by which the research centers were to accomplish such allocation was to be acquired through collaborative relationships between academic and industrial participants. But collaboration made the centers vulnerable, on the one hand, to the pressures of some business affiliates who wanted to use the center for cut-rate contract research and, on the other hand, to the contrary pressure of other corporate affiliates and faculty members who wanted to maintain professorial control over the research agenda. The research agenda in each center emerged from a resolution of this tension. The centers never formulated a means by which they could allocate research projects according to a more coherent notion of the effects of photonics.

Finally, in the form of privatization called "sheltering," examined in chapter 7, R & D programs directed at photonics and intended for industrial effect took place under the auspices of the military establishment. With well-developed technological forecasting abilities in the armed services, the military agencies could have been the ones to respond most coherently to initiatives intended to have them take a role in rescuing the civilian economy. Indeed, such a role had merit in military terms, since photonics had extensive military applications, probably more so than any other contemporary technology aside from electronic computing. Domestic suppliers for such a strategic technology were seen as essential to military preparedness.

Various military agencies pursued policies directed at specific industrial sectors to increase productivity, strengthen the ability to mobilize for war, or maintain viability in the face of foreign competition. Such programs could gain strong political support. Sheltered in obscure military bureaus, the programs managed both to serve the constituencies that support military appropriations and to mollify those opposed to explicit forms of economic intervention.

But military agencies responded with ambivalence to calls for their involvement in industrial policy. Programs oriented toward the civilian industrial technology base conflicted with cherished military priorities. In response to military worries and external political pressure, the agencies did set up disconnected research programs meant partly to strengthen civilian industy, but they never integrated them with the services' own technological priorities. And the agencies always had the incentive to simply make claims on industrial spinoffs and economic effects, without investing serious institutional effort toward such ends, since they would in any case not be formally taken to task for results. In military industrial policy, as in other forms of U.S. response to the rise of photonics, technological research programs emerged as a clutter of discrete choices made in the absence of broader vision and planning.

IN ANTICIPATION OF CONCLUSIONS

In the three forms of privatization, therefore, public decisions on technological priorities were driven to the nether worlds of budgetary negotiations, to public-private collaboration, and to the shadow of the military agencies. Each form of privatization failed by the very criterion that necessitates industrial policy: each failed to respond to photonics as a common resource in industrial production. The sum total of policy responses to the rise of photonics represented, in effect, an industrial policy targeted at photonics. But it turned out to be a debilitating industrial policy in which decisions on public ends were made without strategy, vision, planning, or public debate.

Chapter 8 concludes that photonics does not appear to be unique in the response it has elicited. The U.S. response to biotechnology and semiconductor technology also seems to have taken privatized forms. More generally, other collective resources, such as physical infrastructure and skilled knowledge in the labor force, seem to have integral characteristics, as technologies do. And urban land-use and job-training policy are similarly carried out through privatized policy-making bodies. Privatization, then, may well represent the characteristic means through which the United States allocates common industrial resources. Such a debilitating inability to plan bodes poorly in an era when U.S. industry must adjust to a world economy. Yet these privatized structures (and collaborative groups in particular) might themselves provide the institutional foundations for a more coherent and accountable U.S. industrial policy.

TWO

The Rise of Photonic Technology

PHOTONICS: DEFINITION, APPLICATIONS, AND SYSTEMS

During the 1980s, as photonics was incorporated into the products, processes, and services of disparate industries, the technology came to affect the fortunes of numerous firms. In the United States, however, the spread of photonics yielded another of the decade's examples of our industrial decline in comparison to other industrialized nations. The present chapter documents this development, leading us to ask how we can explain this lagging performance.

To raise the question, this chapter has the initial purpose of showing that photonics has become extraordinarily important—important enough that the ineffectual U.S. response to it warrants an explanation. To this end, the next section of the chapter traces some of the history of photonics. It shows how the technology has its origins in technical developments that appeared early in the century in classical optics and photography and in recent decades in lasers, fiber optics, and optoelectronic image capture. Contrary to a popular preconception, the technology came into being more through the work of the military establishment and intelligence agencies, foreign and domestic telecommunications monopolies, and the universities that these organizations funded than through the efforts of entrepreneurs and private firms. The predominant role of giant nonmarket institutions in the rise of photonics suggests that the technology emerged through the purposeful strategies of these organizations and less through an aggregation of the self-interested choices of market actors.

The section that follows provides quantitative indicators of the rapid growth and present magnitude of the industrial applications of photonics, while also giving evidence of lagging U.S. performance. This country lost ground despite domestic investment that equalled or exceeded that in the country that became the most successful in the field, Japan. Moreover, the Japanese success occurred in the company of a determined government strategy. This observation starts us on a quest for an explanation.

But first, it is necessary to understand what photonics is and the nature of its technical applications, as well as some inescapable technical vocabulary. (For a more substantial technical survey, readers can turn to a number of works that present the new technical field to nonspecialists.[1])

Toward a Definition

As used in its present sense, "la photonique" seems to have made an early appearance in a 1973 article by the French physicist Pierre Aigrain. "I believe," he wrote, "that tomorrow, that is to say in 1990, photonics will play an important part in the transmission of information . . . Photonics is a technology of tomorrow."[2] Also in the present sense, the word *photonics* begins to appear in print in English around 1981 in press releases and annual reports of Bell Laboratories, in internal publications of Hughes Aircraft Corp., and in the more general press in an article in the *Wall Street Journal*.[3] In something of a milestone, the trade magazine *Optical Spectra* changed its name to *Photonics Spectra* in January 1982.

Photonics can be defined as the engineering applications of light, where light is understood in its technical sense as the visible range of the electromagnetic spectrum from red to violet, along with nearby infrared and ultraviolet wavelengths.[4] Photonics uses light to detect, transmit, store, and process information; to capture and display images; and to generate energy. The technology and its subfields are also called by several other hyphenated words and neologisms, each of differing emphasis and unclear definition. Among them are *optoelectronics, lightwave technology,* and *advanced optics*. Still others are *electro-optics* and *optronics*, terms favored in military circles. *Optics* is closest in covering the same breadth of phenomena. But *optics* brings to mind a traditional art of manipulating light through arrangements of lenses; the new word *photonics* connotes the startling newness of the technology itself.

In photonic devices and systems, photons perform many of the functions once performed by electrons; consequently the new technology is often explained by analogy with electronics. To the aficionados of photonics, the analogy seems quite apt, since they expect the technical possibilities to be as revolutionary as those of electronics.[5] Where the technology is well-enough developed, photons outperform electrons. The photons travel at the speed of light and can, by reason of their speed, carry enormous amounts of information at low cost. (Electrons are slowed down by the resistance of the conducting material and can travel at light speed only in superconducting materials or at extremely low temperatures.) They can pass through each other with minimal interference, potentially allowing for greater miniaturization than even microelectronics has accomplished. They can travel long distances without reamplification and are resistant to electromagnetic interference.

Light, unlike electricity, can capture images and is, of course, essential in their display. Thus many recent technical applications termed "imaging" depend on photonic technology. By means of the laser, photonics also has high-energy applications in surgery, welding, precision cutting, and weapons. But the low-energy applications of photonics in information processing and imaging have become economically far more important. After a brief review of directed energy applications in such uses as welding and surgery, the more economically important low-energy applications will be described.

Directed Energy

Applications of photonics for cutting, welding, or destroying are dependent on the laser, a device producing light with characteristics that remain intriguing, even though the devise has been in existence for three decades.[6] Conventional sources of light, such as a light bulb or the sun, emit light that mixes a wide range of wavelengths and tends to disperse. A laser emits light waves simultaneously and in step with one another.

The beam of laser light is very intense or bright, can be attuned to a precise frequency (color), and can remain thin when projected for long distances. Such properties allow it to be used for cutting, welding, surgery, and experiments on nuclear fusion. It can also function as a weapon. In proposed laser defenses against missiles and warheads, the laser beam could damage such projectiles by imparting heat or a powerful pulse. The heat damages the missile's skin or electronic components, while the pulse produces mechanical shock waves that disable the object.[7]

These scientific, medical, industrial, and military uses support a high-energy laser industry of moderate economic value.[8] Far more massive in economic effects are the low-energy applications of photonics.

Communications

Applications in communications have also partly depended on the laser, but on a very low-energy device, the semiconductor laser. About the size of a pinhead, the semiconductor laser produces extremely rapid flashes of light, this light having a very narrow beam. When the light is fed into one end of an optical fiber (the fiber often described as being as "thin as a hair" and "resembling gossamer"), the fiber transmits the signal at the speed of light. According to a commonly used example, one fiber optic strand can transmit the contents of the *Encyclopedia Britannica* in one second. One strand can carry 10,000 or more (as the technology develops, the numbers change) telephone conversations, compared to 48 in a copper wire. At the receiving end of the optical fiber, a photodetector retranslates the light signals into electronic information, passing it on to a computer, a telephone, or a printer.

However, the computer or printer that receives the retranslated electronic data, and the electronic equipment that originally feeds signals to the semiconductor laser, cannot process information at anywhere near the speeds at which the the optical fiber can transmit it. Translating electronic impulses to or from light impulses, the optoelectronic equipment at each end of the optical fiber limits the rate at which information can be sent and is also a relatively expensive part of optical communications. Current research is exploring the feasibility of integrated optical circuitry, which would bypass electronic processing, ultimately making possible the light-speed optical computer.

Since the optoelectronic translation is expensive relative to the optical fiber that carries the information, optical communications has been mainly used for long-distance communication, in which there is a low ratio of optoelectronic equipment to fiber optic lines. As the optoelecronic translation equipment becomes cheaper, it will become feasible to bring optical communications into homes and businesses, thereby vastly decreasing the cost of sending information in and out of homes and offices. When this occurs, high-quality voice lines, high-definition television, switched video (in which programs are ordered from a menu), reference libraries, and business information will enter the home, as will other yet unimagined services and nuisances.

To receive such high-density information service, residences and businesses wired with conventional copper will have to be rewired with optoelectronics and optical fiber. It is with such prospects in mind that forecasters of the future of optical communications predict enormous commercial markets.[9]

Information Processing and Storage

Even by comparison to the processing speed of contemporary microelectronics, optical phenomena have much to offer. Conventional computers transmit bits of information at a speed constrained by the resistance of the conducting material. The computer's microelectronic circuits are also limited in the density to which they can be packed on one chip; when chips are packed too densely, their electrical signals interfere with each other. The potential for a range of optoelectronic alternatives to conventional silicon semiconductors is motivating corporate and military research in several countries. Corporations, especially Japanese ones, periodically announce improved versions of hybrid optoelectronic chips.[10] Critical in the evolution of the optoelectronic chip is experimentation on a range of exotic hybrid semiconductor materials, of which gallium arsenide is the most well known. Such compound semiconductors sandwich together materials with both optical and electronic properties.

Optoelectronic chips are not yet in significant commercial use, but optics for other computing functions are well at hand. Active development is underway in fiber optics to interconnect conventional semiconductor chips to each other and electronic circuit boards to each other. Fiber optic devices connecting local computer networks are already common. And optical storage devices, which store far more information on a compact disk than magnetic computer memories can hold, have become a common tool in offices and libraries.[11]

The all-optical computer is still far away, but a number of military and university research programs work feverishly on developing them. If perfected, they would outwardly resemble a conventional computer, but could perform far more functions for the size of the box, since the computing would happen at the speed of light. The feasibility of such optical computing is hotly debated among specialists,[12] and the technical outcomes seem to be a decade or two away. Whether through hybrid optoelectronic circuitry or through the optical computer, photonic technologies hold the promise of reducing the costs of computation even further than they have dropped in the past decade. While the results most directly affect the world of computing, they also reinforce the applications of photonics in communications, sensing, and imaging.[13]

Sensing

Since fluctuations in environmental conditions affect the movement of light, a number of fiber-optic and optoelectronic sensors have been developed for monitoring engines and machinery, for tracking industrial processes, and for probing in medical diagnosis. They operate on the principle that changes in temperature, pressure, composition of gases, radiation, or other variables affect the propagation of light and can be measured through optical means. They can operate despite magnetic interference, heat, and inhospitable conditions that can interfere with the workings of other kinds of sensors.

Moreover, information from sets of sensors can be interconnected in optical communications systems that can handle vast amounts of information. Systems of sensors, fiber-optic communications, optical memories, and eventually, optoelectronic circuitry, can be envisioned for monitoring engines, machinery, ships' holds, industrial processes, and, in the concept of the "smart skin," the signals reaching an airplane's exterior.[14]

Imaging

The field of imaging includes conventional photography, in which light passes through a system of lenses to leave its impression on sensitized materials like photographic film. The rubric of imaging is more often applied, however, not to chemical photography but to devices used in electronic (more precisely, optoelectronic) imaging to capture, store, display, and print an image. Though conventional digital computers are central to electronic imaging, they operate as part of larger systems that use devices loosely grouped as "electro-optics" for image capture, storage, and display.

While some image-capture devices react to light reflected from a naturally occurring scene, others scan and digitize exposed photographic film. The means most commonly used for capturing the image is the charge-coupled device. Light falls on the device's grid of photoreceptors. These convert light into electrical impulses, which divide the image into picture elements, or pixels.

Once the image is captured, a computer represents the image as discrete points, each of defined brightness. Computer programs alter the image by enhancing contrast and detail, filtering out unwanted detail, moving elements from one image to another, zooming, panning, and changing colors. Programs can also take measurements of elements in the image. When processed, the image can be stored on optical disks, or other storage devices for later retrieval.[15]

The image can be observed on any of several kinds of displays. Since even the most common kind of display technology, the cathode ray tube used in television sets, produces images that must be viewed, it can be considered an optoelectronic device, even through the cathode ray recreates the image through a beam of electrons. Much current research on flat-panel displays is exploring methods of using beams of light, such as laser beams, to inscribe images onto the panel.[16]

Though in the 1980s devices became available off the shelf to perform any imaging function, optoelectronic imaging still had technological limitations. One technical frailty was that image-capture devices, such as charge-coupled devices, offered only resolution. The best commercially available devices as of 1986 could represent pictures as about four hundred thousand points of information, while a good photographic slide contained up to eighteen million. Some reports claimed that this gap was already being bridged by 1990. It was a second shortcoming that the need to perform computations on that many points of information for each picture made enormous use of computer processing abilities. It was less the development of imaging devices, than the ongoing reduction in the costs of computer processing, that caused excitement in the business world about the prospects of electronic imaging. Electronic imaging was seen by many as the most lucrative of the next generation of computer applications.[17]

Photonic Systems

In most of its low-energy applications, photonics is put to use not in a discrete piece of equipment, but in a system that brings together multiple pieces of equipment and software through protocols and standards of interconnectivity. Photonics has its widespread impact on the economic world through technologies that are systematically interrelated.

Imaging applications provide just one example. The speed, flexibility, and low cost make optoelectronic imaging appealing for numerous purposes. Optoelectronic imaging is valuable in aerial warfare, during which an airplane's computer digests masses of pictorial data on ground surface features, targets, and enemy aircraft. Imaging also serves organizations dealing with large numbers of photographic documents and many simultaneous users. These organizations include hospitals, meteorological and mapping services, insurance firms, publishers, and intelligence agencies. Corporate headquarters coordinating numerous branches and medical institutions hoping for nationwide transmission of medical records can also turn to optoelectronic imaging systems.

Such imaging becomes more potent when users can view results on high-definition monitors and even more powerful when pictorial data can be transmitted along telecommunications lines. During the 1980s, the proliferation of fax machines demonstrated the demand for the transmission of images. Through fiber-optic lines reaching into homes and businnesses, far more voluminous and detailed pictorial information will be tramsmitted at low cost.[18] And the entry, editing, copying, viewing, storage, and transmission of images will no longer have to be conducted through separate pieces of equipment, but will be performed through integrated optoelectronic systems.

Since the transmission of imagery requires the processing of immense volumes of information, such developments will become more feasible when optoelectronic or optical computers are perfected. One will then be able to rapidly retrieve selected information, manipulate it, analyze it, view it, distribute it to others in different locations, integrate it with other images or with text, all at light speed and minimal cost. Optoelectronic imaging will gain its power, therefore, by incorporation into larger communications and computing systems.

The dependence of photonic applications on large systems suggests an argument pursued in the next chapter, that photonics should be understood as having integral properties. But before going on to this argument, let us examine the origins of the technology.

EPISODES FROM AN INSTITUTIONAL HISTORY OF PHOTONICS

Origins in Traditional Optics and Photography

Photonics emerged in the 1980s at the confluence of several streams of technological research, some having sources in the previous century. After briefly introducing early optics, the selections below give episodes in the history of three of the more important technical progenitors: lasers, fiber optics, and satellite imaging.

Photonics has its most venerable origins in classical optics and in the applications of optics in opthalmic goods and optical instruments. European makers of precision optics in the nineteenth century produced lenses for microscopes, telescopes, surveying, and navigation. By the 1890s, Schott Glass Works, Zeiss's optical workshop, and George Eastman's roll-film company had developed industrial skills that would lead them through a century of commercial success in optics.[19] In the United States, Rochester, New York, became the home of optics. It housed Kodak and Bausch and Lomb and became the site of the establishment of the Optical Society of America in 1916.

By the next year, the combatants of World War I relearned a lesson of the Boer War—artillery had become accurate enough that optical insturments were now essential for aiming and controlling it. But the Allies found themselves deprived of Germany's optical instruments during the war.[20] After the armistice, the Allies established optical-research institutes in London and Paris and, in the United States, at the University of Rochester.[21]

Development of optical technology in the interwar years occurred in corporate research laboratories then being set up, especially in the photographic industry.[22] During the same period, the invention of the television camera brought into being the first optoelectronic technology.[23]

Another world war passed, bringing in its wake technologies that would require more sophisticated optics. By 1955, technical problems in measuring aircraft and missile flights brought together the founders of the Society of Photographic Instrumentation Engineers (SPIE) in 1955.[24] Later renamed the International Society for Optical Engineering, it was to become the organization most closely associated with the many branches of photonics.

Out of the Death Ray: Fragments from the Story of the Laser

Laser technology also originates after World War II in the capabilities gained with radar and in the search for revolutionary new weapons.[25] But the underlying principle in physics goes back to a 1917 paper by Albert Einstein, and one of the inspirations for the technology can be traced even further back, to the 1898 H. G. Wells's story of the Martian death ray.[26] According to one specialist on the military uses of lasers, a search of classified documents of the 1950s would show that the concept of the death ray was indeed a main motivation behind early investment in laser research.[27]

Physicists and engineers learned the lessons that eventually led to the laser while they spent their World War II service in radar and countermeasures laboratories. Among them were Charles Townes of Columbia University and Arthur Schawlow, a research physicist at AT&T Bell Laboratories who became Townes's close collaborator. In 1953, Townes and Schawlow demonstrated a device, the *maser*, that could emit microwaves simultaneously and in step with one another. The device had only specialized scientific applications, but it spurred theoretical work on the possibility of producing similar effects in the optical wavelengths of the electromagnetic spectrum. Eight months' work begun in 1957 led Townes and Schawlow to a theoretical paper on "optical masers." This line of work eventually brought to each of them patents granted in their names and Nobel Prizes.[28] During this same period related breakthroughs occurred in Japan and the Soviet Union.

Their work in 1957 coincided with the Soviet launching of Sputnik. The launch spurred the United States government to increase its spending on military R & D, and to establish the Advanced Research Projects Agency (ARPA). Townes's 1958 proposal to the Air Force Office of Scientific Research was processed in possibly "the shortest time between the receipt of a proposal and the history of AFOSR."[29]

Also in 1958, Gordon Gould, who quit graduate studies at Columbia's physics department, and notarized some notes that would later allow him to successfully contest the early laser patents,[30] decided to pursue his laser research at TRG, a private firm. When the company asked ARPA for $300 thousand to support Gould's research, ARPA answered with a contract for almost $1 million.[31]

According to a historian of laser developments of the period:

> Both as organizers (for example, of committees and conferences) and as funders, scientists in the Department of Defense's research agencies played a catalytic role in advancing those areas of science from which there might be payoff for military technology... In addition, academic and industrial scientists had a close familiarity with military problems, through service during the war, or through the DoD committees, consultantships, and contracts that became commonplace afterward.[32]

At the laboratories of the aerospace contractor Hughes Aircraft, Theodore H. Maiman demonstrated a working laser for the first time in 1960. Twenty years later, one of Maiman's original team members recalled that "in the beginning it wasn't at all clear what the laser might be good for. In fact, the word was out that the laser was a solution in search of a problem."[33]

For the next twenty years, most of the problems that the solution found were military ones. The first military laser range finder was demonstrated in 1962, but it was not until the end of the 1960s that such technologies came to practical fruition.[34] In the words of an editor with long experience in the laser field, "The exciting applications in 1969 and the early 1970s...were 'smart bombs,' which were field tested in Vietnam, rangefinding for tanks and aircraft, simulations for training and troops, and scoring for war games."[35]

The assistant for research to a deputy under secretary of defense describes this turn of events more graphically:

> In 1972, in Operation Linebacker, the Tahn-Hoa Bridge was completely destroyed via a laser designated bomb, after repeated attempts with ordinary bombs failed to eliminate the bridge at a cost of over 100 aircraft damaged or lost. Since then laser technology has truly revolutionized conventional warfare and all services have been able to escalate their fighting performance dramatically.[36]

In the 1970s, laser technology was also being developed by the United States for an ostensibly nonmilitary purpose, research on inertial fusion. This research intended to test the feasibility of controlled fusion by aiming high-energy lasers and particle beams at a small pellet of fuel. Though placed in civilian agencies (such as the U.S. Department of Energy's Defense Programs), the research had direct relevance for nuclear weapons development as well as for fusion energy.[37]

In the 1960s and early 1970s, AT&T seems to have been the primary developer of civilian applications of the laser. A director of the Engineering Research Center at AT&T's manufacturing branch, Western Electric, declared that the company pioneered industrial applications, with a laser to drill holes in diamond dies used for wire drawing. With frequent conferences with their associated researchers at Bell Labs, scientists at Western Electric also found applications in welding, materials evaporation, measurement, inspection, and holographic imaging.[38]

By 1963, twenty to thirty new firms arose for the sale of laser equipment, and some four hundred to five hundred companies were conducting laser research. According to Joan Lisa Bromberg's history of the laser, these were times of great entrepreneurial fervor about the technology. Scientists pursued laser research for professional recognition and for the joy of exploring a fascinating physical phenomenon. R & D managers sought to enhance laboratory reputations and technological capabilities and laser firms' management sought profitable markets.[39] The expected entrepreneurial forces were at work and helped stimulate interest in the technology among other institutions, such as professional engineering societies and military research agencies.

Though entrepreneurship played its part, the primary force for the development of the technology emanated from another source. Already, the laser firms of the early 1960s depended for their markets or for their research funds on military contracts. Founders of Spectra Physics, which would become the first laser firm to be traded on a national exchange, discovered that federal sources, such as the Naval Research Laboratory, were eager to fund unsolicited proposals from well-known researchers wanting to work on a promising laser topics, even if the companies they represented weres new and unproven.[40] Through the 1970s, for nearly all R & D funds and for most of their sales of hardware, laser companies depended on the government.[41]

For 1980, figures happen to be available on laser R & D expenditures by sector. The output of laser goods and services reached just over $1 billion, of which $349 million reflected companies' income from laser R & D. Of laser R & D in that year, 65 percent was military and 33 percent was for other branches of government. Only 2 percent of laser R & D had nongovernmental sponsorship.[42]

The three decades of federal research investment finally proved its worth to the television-watching audience in February 1991, when laser-guided bombs launched from U.S. stealth bombers made their way into Iraqi bunkers through the front doors. The technological prowess demonstrated in the Gulf War should assure the further development of laser technology for decades to come.[43]

The Military, the Corporations, and the Gossamer Strands

Demonstrating a principle that would later be incorporated into fiber optics, the Irish physicist John Tyndall showed in 1854 that, in certain conditions, a beam of light follows a curved stream of water rather than exiting it in a straight line. Building on this principle, scientists in the 1950s showed in theoretical work that coated glass or fiber can transmit images without distortion. Early applications of optical fibers appeared in the 1950s, not for sending coded signals, but in endoscopes used by physicians to peer inside body cavities and in several kinds of light displays, as in automobile dashboards. In such instruments light in the fiber attenuated and dispersed very rapidly, but since the fibers were only a few inches long, these technical losses were acceptable.[44]

The same technical problems, however, prevented the use of optical fibers for transmission over longer distances. Researchers in the 1960s, most notably Charles Kao and G. A. Hockham at Standard Communications Laboratories, demonstrated theoretically that these losses could be reduced.[45] The British Post Office (later known as British Telecom), Bell Labs, and Standard Telecommunications all initiated research to tackle this problem. Researchers in France and Japan also succeeded in reducing losses.

But the telecommunications organizations had little experience with the material on which the practicability of optical communications would hinge. This was glass. According to one account, scientists at the British Post Office mentioned the possibilities of using glass for such purposes to a visiting scientist from Corning Glass Works (later, Corning Inc.). After his return, Corning established a group led by Robert D. Maurer to investigate fiber optics.[46] In another account, the British Post Office and AT&T Bell Labs specifically approached Corning to gain its opinion on the feasibility of a low-loss optical wave guide.[47]

Around the same time, researchers at Corning Glass began the research project that would further reduce attenuation and dispersion of light, and would find economical fabrication techniques by the late 1970s. Since the primary future application was seen as telecommunications, Corning entered into joint-development and cross-licensing agreements with AT&T and telecommunications cable suppliers in Europe and Japan.[48]

In what was to add a critical component to fiber-optic systems, researchers at IBM, the DoD-funded Lincoln Laboratory at MIT, and General Electric almost simultaneously announced in 1962 the production of semiconductor diode lasers, which produce coherent light measured in milliwats. By 1970, researchers at Bell

Labs and, almost at the same time, some in the Soviet Union, created the "heterostructure" form of the semiconductor laser, which was more easily fabricated than previous semiconductor lasers and potentially had a million-hour life. It was this device that would rapidly send flashes of light into the optical fiber.

Further R & D—much of it at Bell Labs—yielded fabrication methods by which fiber could be manufactured economically; the optical repeater, which maintained the strength of the optical signal over long distances; the photodetector, which captured the light pulses at the fiber's destination and reconverted them to electronic signals; and splicing techniques to connect strands of fiber.[49] Taken together, the fiber-optic wave guide with low information-loss, the fabrication process for making it at low cost, the semiconductor laser (and later, the more economical light-emitting diode), the optical repeater, and the photodetector made possible lightwave communication systems.

Working fiber lines were first laid experimentally in the 1970s, and then commercially, at a rapid pace, crisscrossing the United States in the 1980s. Between 1981 and 1985, phone companies installed fiber-optic trunk lines throughout the country. By 1985, the networks were nearly completed, and the value of fiber-optic shipments reached a plateau at $509 million.[50] The most essential technology that made it possible was developed in the labs of Corning Glass Works, but it could be developed only within the larger world of domestic and foreign telecommunications monopolies.

Even during the mid-1970s, when the commercial value of fiber optics was not yet demonstrated, defense contractors began experimenting with military applications. Starting in 1973, Optelecom, a company in Gaithersburg, Maryland, started work on tactical missiles to be guided through fiber-optic links. The early development led to what came to be called the FOG-M, the fiber-optic guided missile.[51] But such early military involvement seems to have been an exception.

Testifying at Senate hearings in 1980, a leading Bell Labs researcher asserted that the key advances in fiber optics had been made mainly in industrial laboratories and with private funding—the still-regulated AT&T being seen as part of the private sector. But, he continued, government-funded basic research contributed the scientific foundations, and almost all scientists in quantum electronics and related fields who worked on optical communications had received training in government-funded research.[52]

Imaging and the Eye in the Sky

The former director of the Institute of Imaging Sciences at New York's Polytechnic University writes that, though there is considerable R & D going on in the commercial applications of electronic imaging, "much of this activity has been spurred by military requirements."[53] A Wall Street analyst who specializes in electronic imaging claims that imaging systems are "moving into commercial markets after a decade and a half of use in the worlds of academia, NASA, and the Department of Defense."[54] However, if one seeks to trace the recent history of imaging, it turns out to be largely inaccessible, because the development of electronic imaging is tied to the evolving needs of the most secret of U.S. intelligence agencies, the National Reconnaissance Office. This is a "black" organization, whose name and existence are officially secret, though occasional articles have been published on it.[55] Established in 1960 to take charge of spy satellites, the agency is said to have the largest budget of any intelligence agency.[56]

From the early days to the present, most of the agency's satellites have taken photographs on spools of high-resolution film, some spools weighing hundreds of pounds. These are ejected from the satellite, recovered in midair over Hawaii, and transported to the National Photographic Interpretation Center in Washington. Once they are given the developed film, intelligence specialists display the images on photogrammetric equipment that helps them search for assigned targets and measure details, such as the length of a cannon. When clarity of image is the main purpose, the agency can call on satellites that can reportedly distinguish six-inch-long objects on the ground.[57]

Since the 1970s, the National Reconnaissance Office has been sending electronic imaging systems into orbit. Until recently, the largest have been in the KH–11 series. They are equipped with a phototransmission system that relays pictures to Mission Ground Site near Washington. If a satellite is over the proper area, analysts can order close-ups of targets almost instantly, and have photographs delivered to the White House within an hour.[58]

The critical enabling technology in such systems is the charge-coupled device, invented at Bell Telephone Laboratories in 1970. Arrays of the devices, each smaller than a postage stamp, collect radiation, including visible radiation, and translate them into picture elements.[59] Because of the limitations of such digital image capture and transmission, the system's speed and flexibility comes at the cost of resolution—the fineness of the grain. The National Reconaissance Organization and its aerospace contractors, are, consequently, developing new satellites that improve on these capabilities.

The new imaging techniques help meet intelligence needs not envisioned when reconnaissance satellites were first designed. Late in the cold war, the satellite agencies had to redirect satellite imaging for conflicts that were, according to one writer on military reconnaissance, "likely to involve the superpowers indirectly through surrogate states."[60] To better respond to crises in unexpected locations, whether Argentina or Afghanistan, reconnaissance had to use advanced imaging technology to become more mobile and flexible. Image processing had to perform "noise reduction, feature enhancement/extraction, object classification, and recognition" instantly—"in real time." The author concludes that these are the very capabilities now being developed not just for the military but for the commercial world.[61]

In the aftermath of the Cold War, the traditional well-marked boundaries of superpower rivalry disappeared. No one could reliably predict when the next fall of a regime, economic crisis, or Balkan or Middle Eastern enmity would set fire to a conflict. A nation intending to police the world would need to invest ever more funds in technologies that would allow it to quickly collect information on locales scattered around the globe.

Documentation is not available to take the story much further than it has been told here. Despite the scant sources, it seems fair to infer that contemporary commercial developments in optoelectronic imaging owe their technical origins to classified work meant for aerial and space espionage.

By the early 1980s, such military, telecommunications, and corporate research had yielded well-developed technologies of lasers, fiber optics, and imaging. Other lines of technical development not described here, notably the three decades of theoretical elaboration in the field of quantum electronics, accompanied and reinforced the developments in applied optics. As related exhaustively in an article by Paul Forman, the reorientation of physics toward quantum electronics occurred primarily through the influence of military funding on scientists' priorities.[62] In quantum electronics, as well as in laser, fiber-optic, and imaging applications, technical progress originated as much or more in the strategies of nonmarket institutions, primarily those of the military establishment and a telecommunications monopoly, than in technical entrepreneurship or private investment. The strategies, purposes, and politics of these giant institutions have set the course that photonic technology has taken. The amounts invested in some subfields of research rather than others, R & D priorities, the number and characteristics of those employed and educated in photonics, and the location and objectives of research laboratories, all reflect the technology's prevailing origins in these institutions. The directions that the technology has taken reflect institutional strategies far more than automatic economic forces.

COMMERCIAL GROWTH AND DECLINING U.S. COMPETENCE

After 1980, copious technical possibilities and commercial potential led numerous firms to seek their fortune in photonics. Entrepreneurs and established companies swarmed to take advantage of the new field. Other than Japanese data showing remarkable growth in optoelectronics production, the information on the growth of the new field must be assembled from sources of varying reliability.

By 1987, the U.S. commercial directory *Photonics Buyers Guide* listed 2,207 domestic corporate units (including some subsidiaries and corporate divisions) producing goods and services associated with photonics. The number had grown to 2,752 by 1990.[63] Most firms in the directory are small ones, but some of the nation's largest corporations also appear. It shows how seriously the new field is taken that the most shining names in the corporate firmament have created divisions, made acquisitions, or entered into joint ventures to establish their presence in photonics. Dozens of examples can be found just in two of the technology's broad applications, telecommunications and imaging.

In fiber-optic communications, the clearly dominant industrial leaders in the United States from the start of the decade have been Corning and AT&T. But by 1986 several other giant corporations joined the field. DuPont, ITT, Allied Signal, Eastman Kodak, IBM, Celanese, among others, each entered into marketing agreements, opened or expanded divisions, or purchased subsidiaries in fiber optics and related components.[64]

In imaging, too, the big firms have been discerning commercial opportunity. In 1987, the Minnesota Mining and Manufacturing

Company reshuffled its subunits to create a division of information and imaging technology.[65] Eastman Kodak set up a business unit specializing in electronic photography, and Xerox established a Custom Systems Division to develop electronic publishing systems.[66] Not least, IBM set up a department of image technologies in its research division.[67]

The great Japanese integrated corporations also bore down on the new optical technologies. Among them were Hitachi, Nippon Electric Company, Olympus Optical, and Matsushita. Recognizing the potentials of photonics technology early in the decade, Japanese firms formed the Optoelectronic Industry and Technology Development Association.

In 1980 the association began collecting data on production of components, equipment, and systems in which optoelectronics plays a predominant role. These data remain the only consistently collected information for tracking the growth of commercial products incorporating photonic technology.

According to the figures, Japanese production of optoelectronic goods increased from $350 million in 1980 to more than $12 billion in 1988 (table 2-1). The increase seems to be much greater in dollars than in yen, since the yen's value in U.S. dollars rose sharply in 1986. Nevertheless, the Japanese optoelectronics association states that, even in terms of yen, production in Japan had an average annual growth of 50 percent through 1987. The organization asserts that this is a rate of growth "which no other industry can claim."[68]

It is symptomatic of the condition of public knowledge about the economy in the United States (and several other Western industrialized states) that information on a technology's industrial effects is poor if obtainable at all. If one wants to find production or sales data for the United States, one must contend with estimates, guesses, and counterguesses by trade magazines, market research firms, and consultants of varying dependability. In the United Kingdom, the Advisory Council on Science and Technology did set out to estimate the effects of optoelectronics on some of the world's more directly affected economic sectors. The council's rough estimates are summarized in table 2-2.

The least controversial conclusion one can draw from such data is that commercial markets are already enormous. From disparate sources, one can also put together the somewhat safe but not very precise estimate that by 1989 the photonics value of commercial sales worldwide was in the range of $15-20 billion.[69]

2-1
Japanese Production in Optoelectronics

	Japan Billion Yen	Exchange Rate	U.S. Billion Dollars
1980	80	226.6	0.35
1981	160	220.6	0.73
1982	280	249.1	1.12
1983	467	237.6	1.97
1984	642	237.5	2.70
1985	848	238.5	3.56
1986	1,042	168.4	6.19
1987	1,228	148.4	8.27
1988*	1,606	128.2	12.5

* Data exclude new products added to the association's survey in 1988. If they are included, Japanese optoelectronic production in 1988 stands, at 2,033 billion yen.

Sources: Adapted from Optoelectronic Industry and Technology Development Association, Annual Reports for 1987 and 1988. Exchange rates from *Statistical Abstracts of the United States, 1990*.

To gain a sense of magnitude, compare the market value of production derived from another revolutionary technological development, biotechnology. The market for products primarily dependent on biotechnology reached no more than $3 billion in 1989.[70] Biotechnology raises more excruciating moral questions and has greater implications for human health than do most technologies, so it may well deserve the greater public attention it has received. Certainly, photonics yields no trans-genic mice. But it does seem to surpass biotechnology in commercial importance.

Industrial Transformation, Japan and the U.S.

In the 1980s, optoelectronics played a part in Japan's renowned industrial ascendance through exports. According to Japan's Optoelectronic Industry and Technology Development Association, six industrial countries (including Japan) bought $4.8 billion worth of optoelectronic products in 1986. Japan produced $2.7 billion, or 55 percent, of those products. Of nine categories of optoelectronic products recognized by the association, Japan's production exceeded domestic demand in five of those categories and equalled domestic demand in the remainder. For example, Japan

2-2
Sectors to be Affected by Optoelectronic Technology
Estimates of the Advisory Council on Science and Technology (U.K.)

Sector to be Affected	"Market Value" of Sector	Optoelectronic Content	"Market Value" of Optoelectronic Content
Telecommunication Equipment	$147 billion in world in 1986*	Fiber optics	Potentially $1,920 million in Europe; $500 million in U.S. in 1985; $2 billion optoelectronic communications production in Japan in 1985
Information Systems	Over $147 billion in world in 1986*	Fiber optic local area networks, optical interconnects, CD-ROM players, flat-panel displays, bar-code scanning, optical processing, laser printing	"Tens of billions of pounds a year by 1995"
Consumer Electronics	$31 billion in 1986 worldwide	Mainly HDTV and CD-ROM players	Not estimated
Military	Not estimated	Fiber optics, night vision, infra-red imaging, information systems, range finding	$3.2 billion in 1987 in Western world
Automotive Electronics	$28.4 billion in U.S., Japan, & in Europe in 1990	Complex displays, laser-optic wiring, optical sensors	$1.4 billion in U.S., Japan, & W. Europe in 1990

Aerospace Avionics	Not estimated	Optoelectronic devices and systems, cockpit displays, imaging, sensors; information transmission, processing & storage; transparent apertures	Not estimated
Medical Equipment	$30 billion in world in 1985	Noninvasive imaging, lasers, sensors, automated microscopy	$1 billion worldwide in 1986
Materials Processing	Over $294 million	Laser equipment	$97 million worldwide in 1986
Industrial Process Control	$13.7 billion in world	Optoelectronic sensors, detectors, displays, and systems	$32 million in world growing to $150-320 million in ten years
Safety & Security	Not estimated	Light emitting diodes, lasers, sensors, fiber optics, imaging devices	Not estimated
Energy	Not estimated	Photovoltaics	$44 million worldwide in 1986

Note: Where the text does not indicate the year for which an estimate is made, the year 1986 is assumed. U.S. dollar figures reflect conversions, at the rate of US $1.47 = 1 pound for 1986 and US $1.64 = 1 pound for 1987.

Source: Adapted from Advisory Council on Science and Technology, *Optoelectronics: Building on Our Investment* (London: Her Majesty's Stationery Office, 1988).

produced compact-disk equipment worth $1.14 billion in 1986; domestic demand in Japan for this equipment was only $379 million. In that same year it produced $303 million of optical transmitters, of which $287 million went for Japanese domestic sales.[71]

In each of its annual reports, the association continued to project rapid increases in Japanese production of optoelectronic equipment and components, even while excluding the possibility that optical communications would significantly extend local service to businesses and homes—a development that would vastly increase the importance of optoelectronics in the world economy.[72] In view of Japan's history in this field in the past decade, the projections did not seem overly sanguine.

Turning to the United States, one finds a more indifferent industrial performance, though the dearth of reliable data makes assessment difficult. To the extent that it can be ascertained, the U.S. trade balance has generally been favorable in fiber-optic cable, but not in other products incorporating photonic technology. Information on fiber-optic cable is more readily available, because it represents a recognized classification under U.S. customs regulations. (Efforts were underway in 1990 to revise tariff codes to include other optoelectronic products.) Reviewing the information, the U.S. International Trade Commission states that this country experienced a trade deficit in optical cable in the years 1984 through 1986. By 1986, the deficit significantly declined, and by 1987 there was a trade surplus, such changes being partly attributable to the declining value of the dollar.[73]

Fiber-optic cable might have been be the only commercial optoelectronics industry in which the U.S. was doing well. A recent expert group to draw that conclusion was the National Research Council's Panel on Photonics Science and Technology Assessment, which reported on the U.S. position in the technology in 1988. Headed by John R. Whinnery of the University of California, Berkeley, and composed of twelve prominent scientists and engineers, the panel examined the industrial possibilities emerging from photonics and commented on the country's standing in the technology. According to the panel, the largest number of long-life, low-cost semiconductor lasers (which were essential in compact disks and many other photonic products) were produced in Japan. The panel's report added that "the only major counterexample in the photonics field where the United States continues to play a dominant role is optical fibers."[74] A report by the International Trade Commission also suggested pessimism and poor prospects in domestic industries producing optoelectronic components other than fiber for optical communications.[75]

Even in lasers and precision optics, small and specialized industries long dominated by the United States, this country saw its share of the market decline in the 1980s. In 1982, the U.S. held 75 percent of the world market in lasers, while Europe and Japan each held 10 percent. By 1987, the U.S. share had fallen to 38 percent, while Europe's increased to 30 percent and Japan's to 27 percent.[76] In the specialized field of laser optics (making such items as the precise lenses used to focus laser beams), the U.S. share of the world market fell from 62.5 percent in 1984 to 45 percent in 1988.[77] Such figures made it clear that by the end of the 1980s, the locus of industrial strength in photonic technology shifted away from the United States, mostly to Japan.

The growth of the Japanese optoelectronic industries accompanied growing technical accomplishment in the field, while the United States seemed to go through a technical decline. The National Research Council's photonics panel concluded that the United States had become a follower "or worse, an observer" in taking commercial advantage of photonics.[78]

The photonics panel reached this conclusion by comparing the sophistication of recent U.S. and Japanese technical innovations in photonics. Other sets of evidence affirm their conclusion. These include U.S. patent data, data on Japanese technical publications, and the assessments of other panels of experts.

For most of the 1980s, the U.S. Patent and Trademark Office produced a data base that organized patent data in familiar technological groupings, such as semiconductor technology and genetic engineering. These groupings included "light wave communications," a heading that included lasers, light sources and detectors, light-transmitting fiber, and related technology. For each grouping, data were available on the percentage of patents granted that had a foreign origin.

In 1987, just over half of the U.S. patents in light-wave communications had a foreign origin, compared to 21 percent in 1973. The intervening years show an almost inexorable rise in the foreign shares of U.S. patenting. An ever-increasing proportion of the light-wave patents, a proportion reaching 24 percent in 1987, went to Japan. West Germany and France also considerably increased their shares of U.S. patents over the same period (table 2-3).[79]

2-3
U.S. Patents Granted in Lightwave Technology, with Percentages of Foreign Origin, 1963-87

Year	Total Patents Granted	% of Foreign Origin	% Japan	% West Germany	% France	% U.K.
1963-72	1775	17	6	4	2	3
1973	284	21	8	4	4	3
1974	291	29	15	7	4	2
1975	307	32	15	5	3	6
1976	332	38	12	6	5	8
1977	306	41	12	9	7	6
1978	358	36	10	8	4	8
1979	279	44	11	9	8	9
1980	398	43	9	8	11	11
1981	383	45	14	8	8	5
1982	301	53	20	11	7	8
1983	378	49	18	11	8	5
1984	439	46	17	8	6	5
1985	479	48	20	10	6	5
1986	497	48	24	8	6	5
1987	860	53	24	8	5	6

Source: *Technology Assessment and Forecast Reports*, U.S. Patent and Trademark Office

A separate governmental source, a panel commissioned by the U.S. Department of Commerce, warned in 1985 of increasing Japanese technical strength in optoelectronics. After a visit to Japan, the so-called JTech (Japanese technology) panel found the Japanese to be "aggressive in acquiring, improving, and implementing" technologies of American conceptual origin. But the panel added that "In opto-electronics in particular, the Japanese have made major, original contributions and... their original contributions to this field are expected to increase steadily in the future."[80] One of the JTech panel members also traced the number of technical papers published by Japan and the United States in the fields of semiconductor lasers, light-emitting diodes, and integrated optics. He found consistent increases in the number of Japanese papers relative to American ones, although in absolute numbers more American papers were published each year through 1983, the final year for which data were presented.[81]

Still more evidence comes from a special report in a 1986 issue of *Fortune*. The magazine asked ten technical experts to compare U.S., Japanese, Western European, and Soviet standing in four technological fields: computer technology, biotechnology, advanced materials, and optoelectronics. The experts ranked the countries' standing in each technical field on a ten-point scale. The U.S. led the others in computers, biotechnology, and new materials. In optoelectronics, however, Japan was the leader by a substantial margin. "Everyone concedes," the article said, "the Japanese lead the world hands down" in optoelectronics.[82]

Research Investment: Japan and the U.S.

Since skill and knowledge in photonic technology play so central a part in a range of industries producing optoelectronic products, one would expect that investment in research and development would have much to do with a nation's industrial success in this field. It would seem to follow that places that conduct research in a technical field can better take commercial advantage of the new field. But reasonably reliable documentary sources suggest that the faltering U.S. industrial strength in photonics is not attributable to the lack of R & D funding. Japan has been much more successful at developing its optoelectronics industries, despite what seems to be equal or greater research investment by the United States.

Gregory Tassey of the National Bureau of Standards estimated U.S., Japanese, and Eurpean optoelectronics R & D in a report published in 1985.[83] His figures show that Japanese R & D expenditures in optoelectronics moderately exceeded those in the U.S. (table 2-4).

However, Tassey's data omit important categories of U.S. research. The data on government optoelectronic research seem to include only civilian agencies, not the various military agencies funding or conducting optoelectronics research. The survey also does not cover research in universities (which amounts to 9 percent of the total U.S. R & D in all fields in 1986) and in nonprofit organizations.[84]

Data on military optoelectronics R & D are available to supplement Tassey's figures, though the two sets of figures cannot be made fully consistent. One reason for caution in integrating these figures is that military R & D data refer to expenditures that may be performed in government agencies or under contract at corporate, university, or nonprofit labs. Such qualifications kept in mind, let us look at the evidence on military R & D in optoelectronics.

2-4
Estimates for Optoelectronic R & D Expenditures, 1981-85
(in U.S. $ millions)

	1981	1982	1983	1984	1985	1986
U.S.						
Industry	65	98	122	192	250	312
Government*	4	5	6	18	27	27
Total	69	103	128	210	277	339
Japan						
Industry	106	132	160	206	258	322
Government	6	15	21	24	26	22
Total	112	147	181	230	284	344
Europe						
Industry & Government	30	45	65	90	140	165

* Includes only civilian government expenditure

Source: Gregory Tassey, *Technology Assessment of Optoelectronics*, Planning Report No. 23 (Washington, D.C.: U.S. Department of Commerce, National Bureau of Standards, October 1985).

According to estimates compiled from DoD data, total electro-optics research expenditures in 1986 in DoD agencies amounted to $230 million.[85] This figure excludes expenditures by the Defense Advanced Research Projects Agency (DARPA) and the Strategic Defense Initiative Organization (SDIO), which would significantly raise the total.[86]

Just in the field of optical computing, which is a small part of optoelectronics, the DoD data show a $22.7-million investment in optical processing in 1986.[87] It confirms the reliability of the data that a separate inventory of optical computing research in 1986 counts $20 million, an amount that includes DARPA and SDIO but excludes classified projects within so-called black agencies.[88]

Though data are fragmentary, it seems reasonable to conclude that U.S. investment in optoelectronic research and development in 1986 exceeded that in Japan. (Some published information would, however, seem to challenge this conclusion.[89]) As further data will indicate in chapter 7, military agencies have accelerated their spending on optoelectronics research since 1986. In the industries that have grown around photonic technology, the

United States seems to be losing ground to Japan although its R & D investment is greater than Japan's.

The Japanese Counter-Example

Japan has come to lead the world in optoelectronics. In view of the many conflicting claims that have been made about the Japanese economy, one approaches the Japanese counter-example with trepidation. Business journalism, the business self-improvement press, and some academic scholarship have deluged us with investigations and imaginings of what the Japanese do. The literature on Japanese industrial policy is contentious; the role of the Ministry of International Trade and Industry (MITI) is in dispute. Although causes and effects are unresolved, one can safely say that Japan's technological achievement has occurred in the company of deliberate governmental involvement.

Japanese governmental interest in optoelectronics can be traced to 1979, when MITI established the Optoelectronic Joint Research Laboratory as a six-year joint research project among several companies. The next year, the Optoelectronic Industry and Technology Development Association was founded (apparently with MITI backing) with eleven members, a number that has since expanded to several dozen.

By 1985, the joint research project's final year, it engaged about fifty researchers drawn from sixteen companies. Before the researchers returned to their home companies, they undertook a subproject, the Optical Measurement and Control System Large Scale Project, intended to create the momentum by which their companies could start or expand commercial production in optoelectronics. The project used partial government funding to set up an experimental optoelectronic process-control system at a refinery in Okayama. According to U.S. reporting, the project incorporated joint feasibility studies, the design and construction of prototypes, and standardization work.[90]

In 1985, MITI launched a new ten-year project in optoelectronic circuits. MITI's documents portray this not as a stand-alone program, but as part of a broader vision on the evolution of information technology. The optoelectronic integrated circuit program is one of twenty-five joint research projects sponsored by MITI's development program in information technology.[91] Joint research programs operated by other agencies were also underway. The Joint Electrotechnical Laboratory in Tsukuba, the planned science city, conducted extensive research in opotoelectronics. Nippon Telephone and Telegraph set up a consortium of three companies to conduct fiber-optic research.[92] And Kokusa Denshin Denwa, Japan's international telecommunications agency, set up another long-term research program in underwater fiber optics.[93]

As we have seen, by the end of the 1980s Japan built the most successful optoelectronics industry in the world. But the role of MITI in this success and the nature of the interrelations among the Japanese economic agencies are in dispute. The U.S. Congressional Office of Technology Assessment asserts in a 1989 report that MITI and the Japanese telecommunications organizations "initiated a carefully orchestrated campaign" for Japanese domination of the world's fiber-optics and related industries. They "pursued a deliberate strategy of sponsoring domestic industry" by insulating home markets, building a vertically integrated industry, encouraging exports to Europe and the United States, and promoting research.[94]

Chalmers Johnson, who has written a thorough historical study of MITI,[95] writes that the Japanese capacity for making industrial policy does not emanate from any one agency. Rather, it rests on a broader institutional foundation that includes MITI, the Finance Ministry, and other units of the economic bureaucracy. It is also built on the exceptional power and prestige accorded to the men who lead these bureaucracies, Parliament's weak budgetary control of the bureaucracy, government-dominated financial markets, and the postwar consensus that domestic industrial development would have the highest priority. Though he was not writing about optoelectronics, Johnson's work suggests that one should be skeptical of the Office of Technology Assessment's claim that optoelectronic research programs were orchestrated across agencies. He writes that such programs are kept in check through intense competition among the ministries and agencies.[96]

From a smattering of secondhand sources, we cannot make firm assertions about Japanese industrial policy, nor do we need to for present purposes. We need not decide whether agencies worked competitively or in concert. We need only observe that a set of traditions and institutions allowed the Japanese agencies to respond with deliberation to the rise of a technology with revolutionary industrial possibilities. As we shall see, the U.S. response differed sharply from that of Japan.

Conclusion

This chapter has shown that developments in optical, laser, fiber-optic, and imaging technology in the 1980s led to an abundance of new technical possibilities. These converging technological trends marked the rise of photonics. It became a technology with diverse and economically valuable uses. In the commercial value of its applications, photonics became one of the fastest growing industrial technologies of the decade and achieved substantial worldwide commercial importance. It gained this importance in the wake of more than two decades of American private and military investment in research and development in years in which total U.S. investment in this technology exceeded that of any other country. Despite substantial domestic investment, this country's industrial performance in photonics deteriorated relative to international performance.

Whether working in concert or as adversaries, the Japanese agencies seem to have been able to respond with determination to technological change in photonics. The contrasting outcomes in the U.S. and Japan suggest that capitalist nations differ radically in how they take advantage of an emerging technological resource like photonics.

THREE

Technology as a Common Resource

TECHNOLOGY AND INTEGRALITY

The previous chapter showed that U.S. industrial performance in photonics compared poorly to that in other nations, particularly Japan, despite considerable domestic investment. This chapter sets the stage for an explanation, arguing that a massive technological development, like photonics, is not a loose assemblage of innovations. It is, rather, a "technological paradigm." It comprises a cohesive body of knowledge and skill; generates systems of technical devices; and interrelates numerous industrial sectors. Photonics and other technological paradigms exert wide-ranging effects on the economy precisely because they have such integral properties.

Why is the integral character of photonics important in explaining declining U.S. industrial performance in the technology? We can work our way toward the answer by first contrasting to the idea of technological integrality the more usual conception of technology as assemblage of discrete, technical innovations. Indeed, this is the way most of us think of technology: as an accumulation of innovations.

The innovations arise from the disparate researches of scientists in their laboratories, from the work of numerous engineers and technologists building prototypes and testing models, and from the investment decisions of venture capitalists and research managers. Acts of technological innovation reflect their judgments about technical feasibility and their estimations of potential economic value. These are extraordinarily complex decisions. Technological change, such as that represented by the rise of photonics, is then surely an even more complex assemblage of such individual acts of innovation. Innovators themselves, are, therefore, the ones best placed to make decisions on the development of a technology. (Since this atomistic conception of innovation becomes the source of the most important opposition to coherent governmental technology policy, the idea is examined in greater detail in chapter 4.)

But if a technology is understood as having integral properties, private innovators may be unable to respond efficiently to technological complementarities and interdependencies. Technology may then better be seen as a common resource whose development can be directed for common benefit.

In an internationalized economy, where regions and nations are industrial rivals, a capitalist state that can coherently allocate common resources may better enable its private firms to take industrial advantage of those resources. Hence, the short answer to the question posed above follows. In an internationalized economy, in a world driven by technological capitalism, nations may prosper or decline according to their ability to carry out policies that recognize, anticipate, and respond to the integral characteristics of technology.

Thomas Kuhn and the Idea of Paradigms

To elaborate on the idea of integrality, we need to draw on a tradition of scholarship that has examined how sets of shared rules and methods, called "paradigms," have served as the foundations for scientific and technological development. Contemporary thinking about paradigms rests on the work of Thomas Kuhn.

In an exceptionally influential book written in 1962, *The Structure of Scientific Revolutions*, Kuhn argued that scientific advancement should not be understood as a linear progression. It should be seen rather as periods of normal scientific inquiry guided by an established paradigm, such periods interrupted on occasion by revolutionary new conceptions that challenge the old order. The challenges yield new paradigms, which provide explanations for formerly anomalous phenomena and define a new set of problems and methods for scientific inquiry.[1]

The paradigms define the legitimate problems and methods pursued by research practitioners. According to Kuhn, scientific practice (which he takes to include laws, theories, applications, and instrumentation) gives rise to "coherent traditions of scientific research."[2] He writes that

> Men whose research is based on shared paradigms are committed to the same rules and standards for scientific practice. That commitment and the apparent consensus it produces are prerequisites... for the genesis and continuation of a particular research tradition.[3]

The paradigm provides intertwined theoretical and methodological beliefs that enable scientists to select, evaluate, criticize, and interpret data.[4] But the knowledge implicit in the paradigm cannot be reduced to a set of rules. As Kuhn stresses, a paradigm operates partly through tacit knowledge "that is acquired through practice and cannot be articulated explicitly."[5]

By the time Kuhn prepared the second edition of his book, he realized that the notion of paradigms, as he had presented them, posed severe problems of definition and scope. One problem is revealed by this question: What, if anything, is there in scientific progress that does not emerge from paradigms?

In the first edition of his work, Kuhn asserted that there is "pre-paradigmatic" science, which operates when scientists are working with anomalous information outside an established scientific tradition. But if, as his book led readers to think, paradigms are conceptual grounds for scientific acts, then it would seem that a pre-paradigmatic science operates without conceptual grounds. Kuhn rejected this version of paradigms in the postscript to his second edition, suggesting that even those scientists struggling with anomalous information work within paradigms.[6] By expanding the meaning of paradigms, however, he may have compromised the concept's ability to critically distinguish among scientific phenomena. If all science is paradigmatic, then one can no longer differentiate one coherent paradigmatic tradition, like Newtonian science, from another, like Einsteinian science.[7]

His work also left unresolved a second question: Were there other classes of phenomena, besides science, that were driven by paradigms? By restricting examples to science, Kuhn could avoid asking what else might be characterized by paradigms. The problems of definition and scope persisted when a later body of literature sought to expand the idea of "paradigm" to encompass technology.

Technological Paradigms

Recent scholarship has drawn on Kuhn's work to suggest that technological development, like scientific progress, depends on the emergence of paradigms that guide the work of engineers, applied scientists, and other technical decision makers. Moreover, this scholarship has sought to show that such paradigms are essential to massive industrial transformations affecting capitalist economies.[8]

Giovanni Dosi argues that a technological paradigm consists of the concepts, tools, artifacts, and practical experiences needed to solve problems in a technical field. The paradigm gives to technologists particular ways of working in the field as well as general scientific principles. It arises from explicit learning from manuals and schools as well as tacit knowledge (e.g., that of good design and good engineering) learned through training and experience. It reflects knowledge that is private and specific to organizations as well as knowledge that is public and open.[9]

More than a set of rules or concepts, the technological paradigm must also be seen as operating in an interrelationship with physical tools, artifacts, and devices. While the artifacts, whether integrated circuits or lathes, emerge through the application of technological skill and knowledge, they also recursively make conceptual advancement possible.[10]

Dosi explains technological paradigms by offering successive characterizations, while remaining unclear on questions of definition and scope.[11] He does not tell us whether some technologies fall outside paradigms. Nor does he say whether there are any phenomena other than technology (and science) that can also be characterized as paradigmatic.

In keeping with a second tradition from which discussions of technological paradigms are drawn—the tradition of Joseph Schumpeter, discussed below—Dosi sometimes stresses that technological paradigms "become established" as paradigms or otherwise come into the form of paradigms. One has to read Dosi's work carefully to discern what differentiates technological paradigms from technologies in general.

A clarification emerges from one passage in which Dosi contrasts the idea of paradigms with a conventional economic picture that portrays technology as a pool of information that firms can readily use. While accepting that technology does to an extent take on public characteristics because of free-flowing information, Dosi also suggests that technology has collective characteristics for still another reason:

> The second aspect of the 'public' characteristic of technology relates to *untraded interdependencies* between sectors, technologies and firms and takes the form of technological complementarities, 'synergies,' and flow of stimuli and constraints which do not entirely correspond to commodity flows. All of them represent a structured set of technological externalities which can be a *collective asset* of groups of firms/industries within countries/regions. . .[12] (Stress is in the original.)

Though the language is somewhat opaque, one can extract from terms such as *untraded interdependencies,* and *complementarities,* what may be the essential and defining characteristic of paradigms: a set of interrelated and mutually reinforcing developments. While technologists have to recognize interdependence and complementarities, individuals cannot trade them or otherwise adequately respond to them just within markets. By this conception, such interdependence and complementarity are essential to paradigms. Technological paradigms come into being when progress has been made in recognizing and rationalizing such interrelationships.

Technologies can be said to constitute a technological paradigm, therefore, when progress has been made in integrating technical advances, technical knowledge and skills, devices, equipment, standards, and industrial (and military) applications. These interdependencies, complementarities, and other integral relationships constitute the outstanding feature of a technological paradigm, the quality that distinguishes it from technology in general. We can refer to this quality as—in a word—"integrality."

Paradigms and Technological Revolutions

While suggesting, as Kuhn also did, that paradigms operate at several levels of generality and can be observed even in small technological (and scientific) subfields, Dosi and other writers on the subject stress those paradigms that bring about massive technological reorientations in the economy. Their examples are always of revolutionary technologies, ones with widespread effect on the economy. Dosi's examples include the internal-combustion engine, oil-based synthetic chemistry, and semiconductor technology. But he does not state clearly why this level of technology is privileged to receive such attention.

In stressing this, the most comprehensive level at which paradigms operate, some commentators have sought to merge the idea of "paradigm" with the conception that technological changes are the fundamental driving forces in capitalist industrial change. The author to make this connection most explicitly is Christopher Freeman. In accordance with the ideas of Joseph Schumpeter, on whose intellectual tradition he draws, Freeman sees massive technological changes as the very center of the gales of creative destruction that destabilize and restructure capitalist economies.[13]

Yet, once again, problems of definition arise. Freeman, like Dosi, chooses to discuss revolutionary technologies. At one point, Freeman even defines paradigms by their ability to cause technological transformations in nearly every sector of the economy.[14] He uses as examples steam power, electrical power, and the electronic computer. Freeman's argument would be stronger, however, if he did not define paradigms simply by their economic effect but rather attributed the effects to a technology's status as a paradigm.

Indeed, Freeman argues at some points that these technologies do not accrue from a simple bunching of innovations but reflect the emergence of more coherent technological interrelationships. Firms in various sectors must reorient their technical capabilities, supply of technical devices, and corporate strategies to take account of the interrelated effects of a new paradigm. Otherwise, they find that they are overtaken by other firms better adapted to new technological requisites.

Integrality

One can argue that it is the very interrelationship and internal coherence, the integrality, of paradigms that reorient researchers and technological investments to vast new possibilities and affect the fortunes of numerous economic sectors. It is because paradigms are constituted in the integration of technical knowledge, skill, and technical systems that they can generate massively destabilizing technological change.

When integrality is understood as the outstanding and defining quality of paradigms, we can answer a questions posed earlier: Which technologies fall outside paradigms?

Technologies can be said to be nonparadigmatic (or trivially paradigmatic) when they fall at either end of the process by which technical ideas, skills, rules, and devices are integrated into a comprehensive whole. While a new invention or technical idea is still anomalous and has not been incorporated into a broader reconceptualization and reskilling of technological activities, then we may consider it preparadigmatic, even if it is of some economic importance. (To be sure, the anomaly may contribute to the rise of a new paradigm over time.) At the other end of the process, the mature phase of technological change, when the technology has been fully rationalized and routinized, then individual inventors can operate quite independently of any concern about broader transformations in knowledge, skill, or infrastructure.

For example, early electrical inventions, such as the lightning rod, were preparadigmatic. Late in the twentieth century, tinkering to recombine stock electrical components (batteries, capacitors, resistors, wires) into a new invention may be considered postparadigmatic. The paradigmatic phase in the history of electrical technology occurred in between these extremes, running from the late nineteenth century to the mid-twentieth century, when scientific knowledge, instrumentation, physical infrastructure, and industrial production were recognizing and transforming themselves to take account of the technology's emerging integration.

Paradigms are of greatest interest for policymakers, then, at the time when integration is ongoing and problematic, which is the same time that the technology has destabilizing effects on numerous sectors. Therefore, the fundamental reason for privileging revolutionary technologies is that they represent the critical phase in technological integration. As progress is made at rationalizing a set of concepts, skills, devices, and applications on a scale as massive as that suggested by internal combustion and semiconductor technologies, the resulting technological paradigm has destabilizing effects on numerous industries. As this rationalization occurs, research projects, technical investments, physical devices, and conceptual developments come to have integral relationships with one another.

PHOTONICS AS INTEGRAL TECHNOLOGY

In the 1980s, photonics came to exemplify a technology having revolutionary economic effects by reason of its integrality. It came to belong to the class of technological revolutions that presently includes microelectronics and biotechnology and once included petrochemical technology, electricity, and internal combustion.

Indicators of the technology's widespread effect include the rapid growth of industrial applications, educational programs, and the number of engineering professionals who affiliate themselves with the field. But to truly classify photonics as a revolutionary technological paradigm, one must demonstrate its integrality.

The market data given in the previous chapter suggest the rapid growth of the industrial applications of photonics. One indicator of the spread of the new field as a body of knowledge and engineering practice is the burgeoning membership in professional optical societies. The International Society for Optical Engineering has seen its membership escalate from thirty-seven hundred in 1980 to twelve thousand in 1989.[15]

Still another indicator is the growth of research and education programs in the field in the 1980s. Whereas there were only two major university centers for optics research through the 1970s (at the University of Rochester and the University of Arizona at Tucson), the 1980s saw the establishment of more than twenty centers for research in lasers, electro-optics, photonics, and advanced optics.[16] Also during that decade, a handful of new training programs appeared at two-year colleges awarding technical degrees in optics, electro-optics, and laser studies.

Yet, if one suggests that integrality is the essential character of a technological paradigm, then indicators of growth are in themselves inadequate evidence for identifying a paradigm. To demonstrate that photonics, or any other cluster of technologies, qualifies as a paradigm, we should identify its integral properties. In the passages that follow, we will observe photonic technology taking on integral qualities in (1) becoming a cohesive body of knowledge and skill, (2) being put into use through technical systems, and (3) coming to serve as a body of knowledge and practice that interconnects numerous sectors.

Photonics as a Body of Knowledge and Skill

Photonics evolved into a new technological paradigm, first of all, through mutually dependent progress of knowledge and technique on many fronts. Researchers on optoelectronic circuitry depended on others developing compound semiconductor materials. Engineers perfecting optical communication systems required more durable glass fibers. Makers of optoelectronic circuitry and optical fiber relied on developers of the equipment, such as molecular-beam epitaxy and metallo-organic chemical vapor deposition equipment, by which the devices could be fabricated. Each group assumed that still others were working on displays, improved optical storage, and the systems and software that would interconnect them.

Technical progress for all groups depended on the emergence of principles, techniques, instruments, methods, means of calibration, standards, and tacit rules of good practice in the engineering of optoelectronic devices. Engineers working in photonics also learned from and inspired researchers in quantum electronics and optical science who were investigating fundamental properties of electrons and light in various materials. And progress for all concerned depended on a flow of persons newly trained in materials science, optics, and electrical engineering, or in new hybrid optoelectronic engineering fields. As integral relationships arose among these bodies of theory and engineering skill, photonics emerged as a new technological paradigm.

One occasionally sees attempts to express such interrelationships in a diagram. Japanese technologists especially seem to enjoy intricate diagrams, like one depicting a "tree of optoelectronic technology." The schema tries to show the relationships among materials technologies, fabrication techniques, specific devices, various levels at which devices can be integrated, and a number of applications.[17] Another diagram, this one in an American trade publication, shows more linear technological interrelationships in what is labelled a "vertical integration chain."[18] The diagrams are generally intuitive, sometimes whimsical, and almost never seem to have been constructed through systematic investigation. After all, there is no field of technology studies or technology planning through which one could learn how to observe and depict such interrelationships. These occasional diagrams suggest that, through dedicated study, those concerned about technological effects on the economy could conduct rigorous investigations to elicit the paradigmatic structure of a technological field's development.

Whether the technological interconnections are explicit or not, new acts of technical innovation can occur only because they take place amid such mutually interrelated trends. Each new innovative act gains its feasibility in relation to these structured technological developments.

In photonics, since each innovator proceeded from a partial and situated understanding of this new paradigm, no researcher, investor, or firm had the development of the technological paradigm as a whole as its objective. No market actor and no individual organization purposefully created photonics, yet all depended on the emerging field. The photonics paradigm, therefore, became a common resource in the strategies of numerous military, industrial, and academic organizations.

Photonics as Technical System

As converging technological discoveries and concepts yielded the potential for a radically expanded field of technical knowledge and practice, possibilities emerged for potent interconnections among diverse artifacts. With the linking of the semiconductor laser (or light-emitting diode) and optical fibers, excitement suddenly grew in the late 1970s and early 1980s about the possibilities of all-optical switching, optical circuitry and, most intriguing of all, optical computing. As scientists began to envision linking various optoelectronic devices, there was growing enthusiasm for and awareness of the technical power to be gained through the technology's systemic properties.

A technology gains the characteristic of a technical system in the straightforward sense that it is applied through interrelated devices. For example, as described in the work of Thomas Hughes, electrification in the early part of this century linked Western societies in systemic grids that reordered the economy.[19] And the uses of petroleum in motor vehicles and power plants became feasible through systems for refining, distributing, storing, and combusting petroleum for mechanical energy. Though systemic interconnections are more obvious in electronics, telecommunications, and computing, where devices are connected by wires, all broad technological developments are characterized to some extent by such interlinking. When inventions come to be usable in a system of interlinked components and devices, the investment in the development of a device does not affect just the firm or agency that invests in it; it also has consequences for people who make and use the equipment with which it can be connected.

For the purposes of this study, the idea of a 'system' should be understood in a restricted sense to refer to the software and hardware of the interconnections. To distinguish it from other terms referring to the integral characteristics of technology, the word 'system' should be taken to exclude knowledge, research results, civil laws and regulations, organizations, and skill needed to build and maintain the system.[20]

Indeed, applications of photonics gained their technical potency not as discrete devices but in being linked together in networks. Fiber-optic communication linked optoelectronic input devices, optical fibers, optical repeaters, and optoelectronic light-capture devices, all of them dependent on the availability of proper software and communications protocols, materials, and testing equipment. Computing applications of photonics linked processing devices with monitors and optical memories along electronic or optical interconnects. Imaging applications brought together image-capture devices, computers, monitors for viewing images, and printing devices.

Moreover, the power of these applications expanded all the more in their potential for systematic relationships with other classes of photonic technology. Consider only the potential of imaging when it is linked with optical communications and optical or optoelectronic computing. Linked with a fiber-optic communications network, a set of imaging devices can potentially capture, process, print, copy, send, and retrieve images with no geographical limitation. And if economical optoelectronic computing systems become feasible, the cost of processing imagery will decline, further reinforcing the spread to optical communications and imaging systems.

Photonic devices can be used in telecommunications, computing, sensing, and imaging only when they can be integrated into those systems. Users and manufacturers of photonic devices depend, therefore, on standards, protocols, regulations, and the existence of complementary devices. Though firms recognize the necessity of the systems, they cannot individually bring them into existence. This means that, besides having integral characteristics as a body of knowledge and skill, photonics also exhibits its integrality in the form of systems of interconnected devices. Since the users and producers of photonic devices depend on these larger systems but cannot themselves efficiently bring them into existence, the existence of the systems is a common resource.

Photonics as a Technology Interrelated with Industries

As photonics gained its integral characteristics, its impact reverberated in the economy through effects on many and disparate industries. Each of the broad applications of photonics—directed energy, communications, computing and information storage, and imaging—had diverse industrial effects. And as industrial firms (and military organizations) sought uses for photonics, they recursively contributed to its development.

Photonics had integral effects on industrial sectors in two respects. First, as photonics was adopted in numerous sectors, those sectors became interdependent in their reliance on the technology. Second, the technology let loose a set of dynamic effects that destabilized old industries while building new ones.

One can observe the technology's first set of effects simply by noting the range of industries that have become reliant on photonic technology. Photonics is most fully responsible for generating a tier of firms that make lasers, optical components, and materials used by manufacturers of larger equipment and systems. Some are small firms that specifically devote themselves to the production of these technically sophisticated components. Other producers of components are not specialized firms in a narrow industry but giant corporations that consider such production a small part of their operations and do not necessarily sell the components on the open market.[21]

A tier of manufacturing companies makes final products in which photonics plays a large role. They may make machine-vision systems for inventory control and automatic inspection, control systems to run production processes, systems combining fiber-optic probes and imaging equipment for medical diagnosis, image-management systems for voluminous documents; automated fingerprint-identification and personnel-identification systems; equipment for night vision; detectors to indicate trespassers onto restricted property; and optical interconnects for computers.

Giant economic sectors may reorient their operations to embrace such new technical possibilities. The equipment and systems listed above can fulfill crucial technical roles in manufacturing processes, hospital and health services, insurance companies, aeronautics and aerospace industries, police forces, and industrial security services, not to mention military contractors, the computer industry, television, banking, cartography, surveying, retailing, photography, and telecommunications.[22]

These sectors become dependent on the first tier of industries that produce components and devices, while this first tier itself depends on the firms that use those components when they manufacture final products and on the firms that find ways of using those final products to provide services. The interrelationships may take the form of purchases and sales, information exchanges, mutual development of a critical mass of trained persons, and mutual technological learning.[23] Since a business can benefit from such interrelationships in its region but cannot bring them about, the interrelationships serve as a common resource for a region's firms.

One observes the technology's second set of effects in the instability it causes in older industries. Indeed, a technology with industrial effects as broad as those of photonics causes the decline of some industries as well as the creation of new ones.

The growing use of optical cable cuts into the market for copper cable used for long-distance communication, leading to the decline of an old resource-based industry. Optical cable's capacity to carry a high volume of information at low cost also decreases the demand for transmission of information by satellites, thereby undercutting a relatively new industry. And companies making new optical-memory devices for computers are expanding into markets that were once the province of magnetic memories.

Photonics is new enough that such industrial dislocations are mainly incipient. An example of potential dislocating effects can be taken from Rochester, N.Y., the home of photographic optics and, not coincidentally, half of all American workers in photographic equipment and supplies. Through Eastman Kodak and Xerox Corp., the photographic industry provides about 43 percent of the area's manufacturing employment.[24] In the 1980s a range of optoelectronic-imaging devices was introduced, mainly by Japanese firms, at the expense of cameras that create an image by affecting the chemical emulsion on photographic film. The devices were initially intended to have industrial and medical applications, which themselves provide sizable markets for specialized film. Since the new optoelectronic cameras were expensive and offered only low resolution images, traditional consumer photography still seemed secure, and so did Rochester's future, for at least another decade or two. But the technical threat and the potential for regional dislocations were already apparent.[25]

Photonics gives rise to new industries, like optical memories, while destabilizing old ones, like chemical photography. To a greater or lesser extent, it affects nearly every branch of the economy. Individual companies find they must adopt new technologies, produce new products, and respond to new competitors—or risk their demise. In initially investigating a new technology, individual firms themselves do not intend or have the resources to bring about such transformation. At the same time, individual firms and organizations must look to the technology as a shared resource to enable them to respond to the transformation.

In the 1980s, therefore, photonics emerged in several respects as a new paradigm, an interrelated and mutually reinforcing body of knowledge, skills, techniques, standards, and research possibilities. As this body of knowledge and skills was utilized in numerous firms, a web of industries came to be interrelated through mutual dependence on photonic technology. To survive technological rivalries, firms would have to respond to the interrelationship, but could not in themselves bring it about or control it. For private firms, the integral characteristics of the technology constituted it as a common resource.

At the same time, these very characteristics provided the principles by which technological development could have been tracked and anticipated. Characterizing photonics as exhibiting three kinds of economic integrality therefore has implications for policy. The main implication is that, where a nation can use such planning knowledge to formulate policies toward the technology, it increases domestic firms' abilities to succeed amid international technological rivalries.

COMMON RESOURCES IN AN INTERNATIONALIZED ECONOMY

The preceding section has set the stage for viewing photonics as a common resource. The present section argues that some capitalist economies are better than others in allocating these resources. Their private firms' prosperity or decline depends in part on the state's ability to recognize and respond to the development of technological paradigms (and other phenomena that have similarly integral properties). Hence, we might partly explain the deteriorating U.S. industrial position in photonics by looking at the domestic ability, or lack of ability, to respond through policy to the integral properties of technological change.

By this reasoning, the integrality of technological paradigms (and other common resources) warrants technology policy-making in capitalism. In conceiving of such policy, we need to go beyond ideas of "technology transfer" narrowly conceived. The concept of paradigms leads us not to policy proposals for increasing contacts and transfers between institutional sectors, such as university and industry, but to proposals for coherence in industrial technology policy.

To appreciate fully the policy implications of technology as a common resource, we should also ask whether there are common resources other than technological paradigms. This section therefore goes on to suggest that physical infrastructure and training and educational resources, along with technological paradigms, qualify as common resources. The existence of several kinds of common resources in capitalist industrial production suggests an argument for planning under capitalism. This argument supplements other justifications for public intervention in the economy, particularly those that reason from the good and bad side effects of technology. Planning turns out to be valuable even for reasons strictly limited to considerations of industrial competition. The section concludes by suggesting the differences between such resources and public goods.

Integrality and Markets

Technological paradigms warrant coherence and planning in public policy even if we view the capitalist economy—in the simplified and rarified terms of conventional economics—as a competitive market, but one operating in an international economy. Even under such restrictive assumptions, public policy to guide technological development is desirable for two reasons.

First, the integrality we observe in technological paradigms is not simply the outcome of the acts of individual firms competing in their markets. Even though the technology affects the fortunes of numerous firms, its characteristics exceed what we would observe in a simple aggregation of the innovative activities of firms. This observation holds even in the unlikely condition where technology arises only from private firms, without nonmarket institutions. Even in such a radically simplified model, the development of the technology would take on integral characteristics that exceeded the intentions of private firms.

Second, the integral properties of technology give technological development an internal logic that is, in principle, knowable. But while acting competitively in markets, market actors are unable to make efficient use of such knowledge in allocating the technological resource.

For both reasons, a capitalist economy that relies excessively on market actors to allocate technological investments would distribute those investments less efficiently than would another capitalist economy where public policy helped guide technological development. In the absence of an institutional capacity to respond to the integral properties of technological change, market actors operate at a disadvantage, compared to rivals located in a state that does have such capacity. In the state unable to muster a coherent policy response to technological change, the result may be a slow, debilitating, technological and industrial retrogression.

Properly formulated, public policies should develop standards, build research capacities, and make technological investments in response to the integral characteristics of technology. These characteristics of technology may take the form of an ever more rationalized body of knowledge and engineering practice, a system of interconnected devices, and a set of interrelationships among industries. Each of these forms of integrality yields the possibility of knowledge by which public policies can guide technological development for industrial benefit.

With a view to technology as a body of knowledge and skills, the development of the paradigm depends on answers to policy questions over which individual innovators have no control. What are the relative rates of investment in research in the technology's subfields? What is the extent to which developments in the subfields are balanced and mutually reinforcing? Are there critical technical problems that deserve extraordinary investment? In important technical subfields, can capability be purchased abroad as intellectual property or does industrial advantage depend on the cultivation of domestic human resources and research institutions? Are adequate numbers of scientists and engineers being trained? A nation's or region's answers to such questions create the conditions for more effective acts of innovation in resident firms.

With respect to technology as a system of interrelated devices (for example a combined light-wave communications and computing system) the system's evolution depends on answers to policy questions that are not under the control of individual firms.[26] Are proper standards and protocols being formulated? Are compatible components being produced? Is there adequate investment in telecommunications infrastructure? Can the system be designed for the advantage of domestic manufacturers that can produce technologies to be incorporated into the system? If a state is equipped with the means of addressing such questions through public policy, its resident firms gain advantages compared to firms in a state that lacks such abilities.

With respect to the technology in its interrelationships with industries, a nation's ability to take broad industrial advantage of the technology depends on answers to still another set of questions. Are domestic industries available to make use of technological developments in the field? Are certain industries strategically important, pursuing fields of research and application that are needed to complement the activities of still other industries? Is adequate training available for workers who will be expected to apply these technologies? In the United States, no ready data nor any group of trained professionals is available to help make such assessments.[27] Yet the integral properties of technology suggest that there are criteria by which to address such questions. Policy makers could make decisions on public investments (in education and research, for example) by rigorously investigating such technological interrelationships.

A corollary of these observations is that in undertaking industrial technology policy, public institutions should prepare strategies for the developing and shaping integral bodies of knowledge and skill, technical systems, and their industrial interactions. Public policies should attend to those integral technological relationships for which macroscopic policy action is likely to be more efficacious than the incremental acts of individual firms. Technology policies that consist simply of subsidies that return state assets to private firms are, by this logic, flawed.[28]

As the argument has proceeded so far, public policy toward technological paradigms is warranted even when we think of the unadulterated model of a market consisting of multitudinous, freely competing firms. The argument is reinforced when we see capitalism in truer form as the arena for the rivalries of giant institutions, some governmental, some in private hands, and many not easily classified.

As the history of photonics demonstrates, a technological paradigm emerges not only from private investments but also from military agencies and other institutions. The directions of investment in a paradigm such as photonics reflect the politics and strategies of these giant institutions. The level of investment in some research projects rather than others, decisions on who receives education and training, the missions and facilities of research centers—in short, the very directions of technical progress—reflect the outcomes of institutional stratagems as well as the market. The institutional origins of technological paradigms reinforce the integral (as compared to aggregate) character of technological development, since they further lend to technical progress a purpose and coherence that can be known and planned for.

Beyond Technology Transfer

Widespread recognition in the 1980s that technologies frequently owe their origins to nonmarket institutions has yielded a misunderstanding that continues to obfuscate discussions of public policy for industrial advantage. This misunderstanding is found in the idea of technology transfer, in which technologies are usually perceived as discrete items, pieces of intellectual property. Diagnoses of technology transfer as the culprit in U.S. industrial competitiveness have been prominent in the 1980s. They generally posit that U.S. military and academic research has been very successful, but private firms have been unable to gain access to these results, while foreign firms somehow have capitalized on them. This observation has spurred initiatives for industry-university cooperation and for business participation in governmental research, making them the outstanding forms of industrial technology policy in the United States in the 1980s.

Technology transfer would seem to be an important issue in photonics because, as we have seen, commercial applications trace their technological origins to military and other nonmarket institutions. A study by the Office of Technology Assessment, which advises the U.S. Congress on technological issues, observes, however, that in the field of fiber optics very little transfer occurs from government laboratories to private firms or even in the other direction. Security clearances and the onerous procedural requirements of military acquisitions mean that many technology-oriented firms avoid doing business with military laboratories. Where companies do conduct both military and civilian business, they frequently set up separate military and civilian wings, often with separate staffs, laboratories, research results, and administrative systems.[29] The difficulties of direct technology transfer would be compounded in the case of optoelectronic imaging, where much of the work occurs in conjunction with reconnaissance organizations, which are among the most secretive in U.S. government.

But how could we assert, then, that photonics entered the industrial world largely from military origins? Armed with the idea of technological integrality, we can work ourselves out of the apparent dilemma. Indeed, if photonics is properly characterized as a technological paradigm, it follows that the industrial applications need not be traced to any transfer of pieces of intellectual property between institutional settings. Rather, the military-sponsored researches and engineering projects of the 1960s and 1970s produced new bodies of knowledge and skill. These were later incorporated into university engineering curricula, became the themes of numerous conventions of professional scientific and engineering associations, and became the subjects of investigation at new university research centers.

Simultaneously and interactively, scientists and engineers became broadly educated in the new field and applied the knowledge to industrial research and product development. And as the commercial applications of lasers, fiber-optics, and imaging became more apparent, industrial research itself became a prominent contributor to the emergence of photonics.

We can, therefore, trace the origins of photonics in part to the military sector, even if the Office of Technology Assessment is correct in saying that direct transfers of intellectual property rarely occurred. The concept of 'technological paradigm' allows us to avoid seeing the rise of photonics as an example of "technology transfer." The concept allows us, instead, to appreciate that firms, innovators, and investors acquire the ability to make informed decisions about a new technology by participating in an evolving body of knowledge and skill.

When a fundamental technological development arises largely under governmental auspices, private firms, domestic and foreign, can nevertheless participate in it. They can do so even if they did not receive straightforward transfers of intellectual property from military or academic institutions. Once the technology becomes a common resource, it can no longer be appropriated just by the institutions or nations in which it originated. Yet nations and regions that originate technology can, through education, training, and industrial investment, give to their firms timely advantages that can be brought about only more slowly in other places. And nations with the proper educational, research, testing, and standardization programs can gain prompt access to a technological paradigm that has arisen abroad, especially if the country of origin is poorly organized to take advantage of its own common technological resources.

The concept of the paradigm implies that national technological advantage arises less from transfers of intellectual property between institutions than from the ability to plan and guide the development of a technological paradigm for common benefit.

The Technological Paradigm among Common Resources

The idea of integrality also helps us answer another question posed at the beginning of this chapter: Are there economic phenomena besides technology that have characteristics similar to paradigms? The commentators on technological paradigms treat paradigms and sister concepts as unique to technology or, at most, as equivalent to scientific paradigms. If technological paradigms can be said to fall into a broader class of common resources, we can better elucidate the meaning of the paradigm and its implications for policy.

Consider phenomena that are necessary in production and have integral characteristics but, like technology, are also often represented as an aggregate of individual acts. A few examples that share such features are a river valley, an irrigation and drainage system, and a community's educational resources.

A river valley has integral characteristics, as a technological paradigm does. The quality of an aquifer downstream is dependent on the rainfall received and retained in the catchment area upstream. But the river valley, unlike a technological paradigm, is a natural phenomenon and can exist in the absence of human acts. The valley can be classed as a common productive resource in the present sense only when it comes to be diverted, polluted, managed, or used in production. One riparian's use of water resources then impinges on the uses of others; the effectiveness of the valley's use in production might depend on common action in preservation of the valley.

The valley's character as a common resource becomes clearer if we see it as having been engineered into an irrigation and drainage system. Henceforth, the flows of water in and out of the canals depend to a large degree on human acts. Those human acts can occur in the absence of any shared authoritative effort or plan, but such disparate actions are likely to allocate the water poorly and lead to the deterioration of the resource. Development of the irrigation system, like the progress of technology, is constrained by nature. The irrigation system depends on individual human acts, within natural constraints, and also requires a broader authoritative framework for effective allocation. Therefore, the irrigation system (like other forms of infrastructure, such as public transportation and waste management) can be said to fall under the class of common resources.

The attainment of skill and knowledge in a region (i.e., the development of human resources) resembles technology most of all. Similarities and differences can be explored at length, but let us restrict the discussion to the selection of, and the attained depth of learning in, subjects requiring skill. The attainment of skill can easily be viewed as the outcome of individual choices in the search of personal betterment. Where training is given in educational institutions entirely by criteria of demand, skills are implicitly viewed as discrete entities. But skills themselves frequently gain value and recognition within occupations; occupations and professions are institutional rather than aggregate phenomena. And subjects of skilled knowledge evolve in mutual response to each other and through association with science, scholarly investigation, and open debate. Integral relationships arise all the more when skills need to be interrelated with the evolving needs of industrial production. The disposition of skill in the work force might then be better viewed as a common resource.

Therefore, educational resources and public infrastructure, as well as technological paradigms, can be said to fall under the heading of common resources. When an authoritative body applies comprehensive vision to the development of the irrigation system, training program, or technological paradigm, then the choices of the individual water user, learner, and technologist become more efficacious. Put more formally, markets can, in principle, allocate factors of production (land, labor, and technology) in the absence of planning. But they will allocate more efficiently in the presence of public policies that treat productive resources (infrastructure, human resources, technological paradigms) as common resources whose development can be anticipated and planned.

Based on the idea of common productive resources, an argument can be built for the importance of planning within capitalism[30]; on the more specific idea of technological paradigms, an argument can be made for technology policy. These reasons for planning and technology policy should be contrasted with other arguments for state intervention in markets: arguments about side effects and public goods.

Photonics and Side Effects

According to the argument on common resources, the integral relationships of a technology such as photonics in the economy are sufficient to merit a public response. Such reasoning does not exclude—indeed it reinforces—the more usual reasons given for public intervention in markets. The reason most commonly given is that of good and bad side effects. This reason may be given for intervention in photonics as well.

Among bad side effects, environmental ones most readily come to mind. By and large, however, photonics seems to be a relatively benign technology, since it exploits optical processes, rather than chemical or biological ones. But in the production of compound semiconductor materials, which are used in optoelectronic circuitry, the effects may not be at all benign. Researchers and manufacturers have been working with a range of exotic compound materials, whose effects on the environment and public health are not yet known. The material used most commonly, gallium arsenide, appears to have adverse environmental effects.

Arsenic is present in large quantities in the manufacture of this material. Its manufacturing process entails release of toxic vapor and dust and creates serious problems in the disposal of waste products and the circuits themselves. An article warns of environmental disaster from the accidental release of arsine gas and of the dangers of cumulative occupational exposure to toxic substances.[31] If the report is correct, photonics generates not only industrial products and processes but also toxic waste and the chance of environmental harm. Since the private firms do not have to absorb the cost of the harm—the harm is a negative externality in economic terms—governmental intervention is warranted to protect the public.

The technology is also causing excitement in its potential for having applications, especially medical applications, whose benefits are more than commercial. Fiber-optic strands can be made into far thinner and more supple catheters than have previously been feasible. Lasers are increasingly used in surgery that is more precise and less destructive than the scalpel. And a growing array of imaging systems allow for the emergence of a set of less-invasive diagnostic procedures. Since these medical applications of photonics may yield health benefits—positive externalities—on which private investors cannot collect, firms will pay less attention to medical investments than is socially desirable. They will underinvest in the technology from the point of view of the public interest. (Arguments about positive externalities are, as we shall see below, often indistinguishable from arguments about public goods.) Public intervention is called for.

Good and bad side effects are, then, widely understood reasons for public intervention in markets. When such reasons are the sole ones given for state action, the implicit assumption is that firms take care of their own technological needs quite well, and we need to worry only about the social consequences of technology. The argument about common resources stresses, by contrast, that in an internationalized economy, technology policy is desirable even by a strictly industrial logic. Public institutions established for responding to and influencing technological paradigms are needed, even if only to assure proper industrial advantage from technology.

Once such institutions are in place, however, they can take on broader responsibilities of considering the environmental and health effects of the technology. They can potentially guide technological change in the economy.

Public Goods Versus Common Resources

In formal economic thinking, the concept of public goods is used frequently to explain the need for public intervention. How does a public good differ from a common resource? The differences are in (1) the definition of the good or resource, (2) the assumed relationship to production, and (3) the policy implications.

The first difference is in definition. By definition, a technological paradigm such as photonics is a common resource because of its integrality. The resource cannot be effectively allocated if it is understood simply as the aggregate of individual acts. Furthermore, the very integral logic of the technological paradigm (or other common resource) implies that comprehensive characteristics and trends of the resource are knowable. Hence, in an internationalized economy, capitalist states that rely for the allocation of the resource entirely on the situated judgments of individual economic actors will be less able to take industrial advantage of the technology than will capitalist states that also have a macroscopic policy-making capability.

A public good, by contrast, usually refers to a good in which the market invests suboptimally because firms cannot privately appropriate the returns. The social returns to investment in public goods are higher than private returns. The higher social return occurs in technological research—frequently thought of as a public good—because the company that invested in it cannot fully reap the benefits of its investment; others enjoy the benefits without paying. The firm cannot garner all the benefits because patents are insufficient protection, because the very existence of the new product gives other firms signals about what is technically feasible, and because information gained during research diffuses to other firms.

Other firms can use the resulting information to develop products to compete with the firm that did the original research. And the public may benefit from the technology without the original firm being able to reap the financial returns. The other firms and the public are then to be considered free riders. They gain benefits without paying all the costs; they are nonexcludable from the benefits of the investment. Since the firms that wish to invest in research cannot fully collect on the returns, they invest less in such research than is socially desirable.

Besides research, other examples of public goods, such as military defense or a lighthouse, share similarities in principle. Citizens gain the benefits of defense even if they do not pay for it—they are nonexcludable. Ships gain benefits of the beacon even if they do not pay. Hence, the provision of defense and lighthouses depends on the taxing capability of government to ensure that potential free riders pay their share of the cost. To economists, therefore, essential characteristics of a public good are that users are nonexcludable and that social returns are higher than private returns.[32]

A common resource should be understood differently. Research can be a common resource even if patents provide firms with entirely adequate protection. Skilled capability in the work force is a common resource, even if private firms can provide training while using contracts with students to exclude other firms from gaining the benefits of the investment in training. And a public system of navigational aids can be a common resource even if a subscription system can fine ships that do not pay the subscription. Technological research, training programs, and navigational safety systems can each qualify as a common resource because of their integrality. So, while the distinguishing character of a public good is the inappropriability of social returns, the defining character of the common resource is integrality. (If public goods are defined by indivisibility, then the concept becomes similar to that of common resources.)

The second difference between a public good and common resource is in the relationship to production. The theory of public goods generally views the problem as one of consumption, not of production. The public good may even be defined as one that can be enjoyed by additional consumers at no additional cost of production. By underproducing public goods, firms reduce the aggregate welfare of consumers. By contrast again to a public good, the idea of a common resource is specifically necessary in production (though without prejudice to the value of the resource for other purposes, such as environmental protection or public health). Public goods, therefore, are important to aggregate social welfare; common resources are essential to the productive operations of capitalism.

The third difference between the two concepts is in their policy implications. Economists usually draw policy prescriptions from the idea of the public good that are very different from the conclusions we can draw here from the concept of the common resource.

The theory of public goods provides potential tools for determining the optimal size of the governmental budget and the proper division between government and private investment. The theory is also taken to raise questions about means of public choice (such as voting procedures) in the allocation of the good in the absence of an efficient market. The theory does not raise questions about the better locus of knowledge on the allocation of the good. Indeed, judgments on the development and allocation of the good are assumed to be unproblematic; only the preferences of the consumers are in question.

In the idea of a common resource, by contrast, it is knowledge that is problematic, whether knowledge of the proper research investment or of the educational curriculum. While the notion of public goods asks us to look at government budgets or the means of expressing preferences, the idea of common resources asks us to appreciate the means by which knowledge is acquired and policy decisions are made.

Common Resources and U.S. Industrial Decline

Among the numerous works that expound on the problem of U.S. competitiveness in the 1980s, a few give common resources—going by several names—a significant role in their explanations. But the studies offer the explanation only in passing, without critically examining the policy implications that such resources hold for the economy. Moreover, they take one of two tacks in their works, treating the resource either as a set of inter-linked factors of production or as a public good.

In the work of Cohen and Zysman, declines in U.S. manufacturing industries are seen as endangering the skills, technologies, and other capabilities on which many other industries depend. They write that "several industries may depend on a common understanding of production and a similar set of skills for sustaining it,"[33] and that "some critical technologies can affect the competitive position of a whole range of industries."[34] How can government policy help build national industrial advantages?

It can do so principally by substantially upgrading the quality of what goes into production, the factors of production—raw materials, capital, labor—and the networks and rules that affect how those factors are combined—the economic infrastructure.[35]

By accepting the language of factors of production and linkages among the factors, terms that inherently assume discrete relationships among discrete items, Cohen and Zysman and other commentators, though they support coherent industrial policy, leave themselves open to easy rebuttal. This rebuttal focuses on the superior efficiency of competitive firms in allocating productive factors.

Works by William Ouchi and by a team of M.I.T. economists emphasize the role of collective endowments in U.S. industrial standing. "Social endowments," Ouchi writes, "are those public goods that no one company can produce on its own, such as better trained crafts people or better use of land."[36] To Ouchi, the solution to the poor utilization of common endowments is teamwork among corporations.

From studies of U.S. standing in a number of industries, the MIT Commission on Industrial Productivity, composed of a number of distinguished economists, observes that American productivity suffers because of the "underprovision of such collective goods as joint research and development, standardization, and education and training, which were instrumental in promoting technological innovation and productivity in those same industries in Europe and Japan."[37] They, too, do not elaborate on the properties of these collective goods, apparently holding that the traditional economic theory of public goods accounts for them adequately. The authors assert, furthermore, that the goods are underproduced because of the "absence of horizontal cooperation in the United States,"[38] meaning cooperation among private firms and between firms and government.

Therefore, important diagnoses of industrial decline attribute the problem to shared industrial resources but express these resources as inter-linked factors of production. These works are vulnerable to counterarguments that see the market as the supremely efficient allocator of the human, physical, and technological factors that enter into productive activity.

Other diagnoses treat shared resources as public goods whose allocation would be improved through collaboration among firms or between firms and government. Indeed, as we shall see, collaboration has come to be one of the foremost U.S. policy responses to the industrial challenges of the 1980s. But those who call for collaboration fail to ask whether collaborative associations can in themselves respond coherently to the integral properties of those resources on which industrial production depends.

Concepts of linkage and of public goods, then, may well fail to represent the roles of technological paradigms, like photonics, in capitalism. The systematic interconnectivity of photonics, its historical origins in military and public institutions, its emerging character as an integral body of knowledge and skill, and the range of industrial effects to which it has given rise all suggest that it can be understood as the common resource of a nation or region.

Having such integral characteristics, photonic technology cannot be adequately conceptualized as the accumulation of discrete inventions. Having structural integrity, it cannot be disaggregated so as to be traded on financial markets or fought over for political interests, without a loss occurring in the process. And it cannot be adequately understood as a set of discrete factors of production or as a public good. It is better seen as a resource whose effective disposition in the economy depends on policy decisions made outside individual firms. The capability of private firms to produce goods and services efficiently might now depend on the state's ability to plan. The idea of integrality, in contrast to the concepts of linkage or public goods, provides us with an industrial policy argument that stands up against conventional rebuttals.

Because of these same integral properties, a technological paradigm becomes accessible to planning knowledge. In response to an emerging technological paradigm, a nation can, therefore, formulate policies that anticipate technological trends, giving its private firms greater chances of success in international industrial competition.

If this view of technological paradigms is reasonable, then deteriorating U.S. performance in photonics (and generally in technology-intensive industries) may be explained in part by this country's inability to formulate and implement public policies that respond to the integral properties of technological change. Hence the questions that guide the rest of this book: How have U.S. public policies responded to the econimic potentials of a common resource such as photonics? How are we to decide if these policies have mattered?

FOUR

Privatization and the Industrial Policy Debate

THE REJECTION OF EXPLICIT INDUSTRIAL POLICY

The undistinguished domestic industrial record in photonics in the 1980s coincided with a general decline of U.S. industrial performance. That deterioration was seen as being especially disturbing in the high-technology industries. It was in response to the decline that industrial policy had a short life as a collection of proposals for reinvigorating U.S. industry. They included varied suggestions for federal investment, trade, research, and training programs, many having the feature that they could be directed, or "targeted," at specific industrial sectors. To those familiar with the general tenor of public debate in the United States, it will come as no surprise that proposals for explicit industrial policy failed as an item on the political agenda and that the argument for such policy floundered in economic debates. According to the conventional wisdom that became pervasive, industrial policy would be unworkable and foolhardy.

Ostensible rejection of industrial policy despite widespread concern over declining U.S. industrial performance forms the context for the public response to photonics. However, when the predominant public response to photonics is examined more carefully, one finds that it does not consist of the rejection of industrial policy in favor of market-led solutions. Rather, the response in the 1980s was a privatization of policy making. The first purpose of chapter 4 is to introduce the idea of privatization and contrast it to other conceptions of U.S. policy-making.

How are we to evaluate this privatization? Critics of industrial policy would say that such policy was wrongheaded: The idea of industrial policy was misguided at the start, since it hoped to outguess committed businesses in technological decisions. To reach this conclusion, the critics would implicitly assume that technological change occurs through the accumulation of discrete technical innovations, that investments in such innovation would best be left to the self-interested entrepreneurs, investors, and researchers closest to the scene.

But, as has been argued in the preceding chapter, massive technological transformations, like photonics, have an integral character to which individual decision-makers cannot adequately respond. Public institutions are best positioned to formulate the broad visions, plans, and strategies with which a nation or region can respond to such changes. Privatized industrial policy can, therefore, be evaluated according to the following criterion: Has such policy been able to complish the task that would justify it in a capitalist economy, the task of responding with strategy and vision to the emergence of photonics as a common resource? The second purpose of chapter 4 is to argue that the preceding question is the proper one by which we should evaluate privatized policy-making.

To set the stage for a discussion of privatized industrial policy, the rest of the present section examines the intellectual and political context of the industrial-policy debate. It traces the brief history of the debate, showing that a flirtation by the U.S. Democratic party with industrial policy as a political platform brought discredit and embarrassment. The futility of such policy became conventional wisdom just as, curiously enough, sector-specific policies came to be carried out in the U.S. on an ever-broader scale. Operating in intellectual limbo and political disgrace, industrial policy became sublimated. It was transformed into public-private cooperation, goverment-university-industry partnerships, competitiveness policy, defense industrial-base policy, and industry-led policy—the euphemisms of privatization.

A Time of U.S. Industrial Decline

The rise of photonics in the 1980s and this country's mediocre performance in the field corresponded with a more general decline of U.S. hegemony over the world's technology-intensive industries. It was a time of ever-worsening trade balances, the economy was becoming internationalized, technologies of production were rapidly changing, and domestic manufacturing industries were in disarray. The U.S. seemed to be undergoing an industrial and technological retrogression relative to a few other advanced captialist nations. "Competitiveness" gained currency as the economic watchword.

Some of the statistical figures that showed the decline bear repeating, especially the export data. If total exports by the nine major captialist industrial states are traced from year to year, it turns out that the U.S. share of those exports declines progressively, from 29 percent in 1960 to 17 percent in 1987.[1] The absolute amounts of U.S. exports did continue to grow in most years (they could do so since world trade increased overall), but they grew at a much smaller rate than did those of Japan and West Germany.[2] What made matters worse was that domestic demand for imported goods increased much faster than did U.S. exports. Just between 1982 and 1986, imports as a percentage of real domestic demand in the U.S. grew from about 16 percent to 22 percent.[3] The consequence of rapidly rising imports but slowly increasing exports was the infamous trade balance which, after years of dollar devaluations and monthly fluctuations, remained firmly on the multibillion deficit side through 1990.

The competitiveness issue was more than a matter of trade. Domestic industrial plants purchased by foreign firms and operated according to foreign management practices frequently outdid American firms on home territory. In some of the most technologically advanced and commercially promising fields, business initiative and innovation seemed to be shifting abroad. To some, the Unites States was undergoing a historic decline from its position as a great industrial power. For all these reasons, throughout the middle to late 1980s, blue-ribbon committees, industry-government-university forums, and a numbing array of articles in the business press rang the alarms of the competitiveness problem.[4]

What was most unexpected and disturbing to American industrial leaders was that their country was lagging in the research-intensive industries that they had dominated for nearly a half century. Bruce R. Scott and George L. Lodge demonstrated this in a 1985 study of U.S. competitiveness. They found that U.S. shares of exports declined in important high-technology industries: aircraft, computers, instruments, electrical components, drugs and medicines, and industrial components, among others.[5] For example, the American share of the world market in electronics decreased in the 1980s to 10 percent of what it had been in 1965.[6] By 1987, the U.S. had a net trade deficit in several technologically advanced industries, including automobiles, consumer electronics, steel, semiconductors, computers, and photocopiers.[7]

The compact disk player and the videocassette player were the most famous examples, but there were also many other products, some of the most important commercial innovations of the 1980s, in which the U.S. barely got a foothold. In such cases, domestic industries were succumbing not to lower production costs but, apparently, to East Asian and Western European ability to make more effective use of contemporary technology.

Everyone knew that all too many countries were willing to lace shoes and stitch collars for daily wages that wouldn't buy an American a hamburger. But airplanes, semiconductors, advanced consumer products, and computers—these were the industries in which the United States would need to thrive in order to keep its standard of living. They were industries that could be maintained only through advanced skills, productivity, and know-how, despite wages that were high by international standards. To many Americans, these industries were the ones that, by rights, the United States should be doing best in.

A public policy response might have seemed to be warranted. But were the response to take the form of sector-specific policy, an obstacle, possibly an insurmountable one, stood in the way. The concept of industrial policy had all too recently met its political and (in mainstream thinking) intellectual demise.

The Short-Lived Politics of Industrial Policy

By the end of the 1980s, a conception of faltering U.S. industrial performance relative to the rest of the industrial world became commonplace. But a decade earlier, this picture of relative international decline appeared only in incipient form. The picture was muddled by concurrent developments that could not yet be analytically separated. Unemployment, plant closures, decline of the old industrial cities, the ever-lower percentage of the work force in manufacturing jobs, state and municipal fiscal crises, plant movements to offshore sites, stagflation, oil shocks—these were the combined industrial strains of the 1970s and early 1980s.

It was in the setting of this industrial predicament that presidential elections were held in 1980. In the wake of Carter's defeat, factious congressional Democrats flirted with new economic platforms and briefly embraced industrial policy. Edward Kennedy, senator from Massachusetts, expressed the new intention as follows: "historically, the unifying issue for the Democratic party has been the economic issue. We need the restoration of our economy. The basis of that restoration is the development of an industrial policy."[8] And as late as 1983 Walter Mondale saw industrial policy as the blueprint for the Democrats' economic policy for the 1984 elections. He is reported to have drawn this conclusion from the work of Robert Reich, who was then the most prominent spokesman for industrial policy.[9]

The concept of industrial policy differed from the party's earlier economic platforms in that the party implicitly turned away from redistribution, social welfare, and environmental issues and toward economic growth. At the same time, industrial policy held out the hope of reviving the faltering older industries of the Northeast and Midwest and serving the labor and working-class constituencies that Democrats had long cultivated.[10]

No sooner was industrial policy formally adopted by the party than Democrats started backing out. In Congress, hastily conceived programs for a national investment bank and for tripartite committees of labor, industry, and government were never even reported out of committee. Already by 1984, industrial policy played only a small part in Mondale's presidential campaign, and his defeat did nothing to rescue it. Still another death knell for industrial policy was the 1984 defeat in Rhode Island of the Greenhouse Compact, a set of industrial-policy proposals for the state's industrial revitalization.[11]

Why did industrial policy fail as a political platform? One reason was that industrial policy was at first so vaguely and variously defined that factions among the Democrats could stand behind it; but when it was refined into specific proposals, many Democrats balked. In its emphasis on direct public-policy actions, it played poorly against the Republicans' anti-governmental position and came to be seen by Democrats as a public-relations liability.[12] This was also the lesson learned from the Rhode Island industrial-policy proposals, which smacked of elitism and central planning.[13]

Furthermore, by 1984 rapid inflation ended, the baby-boom generation was becoming integrated into the work force, the oil prices had stayed down, service industries provided ever more jobs, and the cities had survived fiscal retrenchment. Dilapidated industrial structures were levelled, the Sun Belt–Frost Belt migration reversed, and the metaphor of a northern Rust Belt lost favor. By the middle 1980s, the sense of a crisis of deindustrialization had been replaced by a tenuous economic complacency.[14]

The Sublimation of Industrial Policy

Industrial policy was not only a political liability. The Democratic party's own intellectuals soon rejected it as faulty economics. And the economics profession then carried out an onslaught on the concept, even though the villain had already faltered. By the middle of the decade, industrial policy was, in its explicit forms, politically dead.

A consequence of the policy's political demise is that it interrupted the dialog about sectoral policies. The opponents of industrial policy made their views well known, but since the explicit proposal was defunct and discredited, few were left to offer a rejoinder.

The premature termination of the industrial policy debate should have been no problem if the economy had simply recovered. But the industrial problems only changed their demeanor. In the midst of heavy governmental borrowing and unchecked consumption, the aggregate statistics did well enough, but industries operating in the world economy, including industries dependent on emerging new technologies, were still deteriorating. U.S. standing in world trade continued to be hotly debated, but in front of a smaller and more elite audience. Commentators now avoided the term *industrial policy* in favor of euphemisms such as *industrial strategy, strategic trade policy, competitiveness policy, government-university-industry cooperation, public-private partnership, defense industrial base and defense technology base policy,* and *industry-led policy.*

With concern widespread among industrial and governmental leaders, it should be no surprise that numerous policies in federal and state government were formulated on the hope of reasserting American industrial standing in the world. One of the primary means of doing so under Ronald Reagan's administration turned out to be initiatives aimed at specific technologies or industries. Among the more visible and large-scale programs were ones directed at biotechnology, superconductivity, semiconductor technology, supercomputing, and the design of a permanent space station, each primarily justified on the argument of competitiveness. As this study will explore extensively, there were also several less-visible initiatives toward a technology that few had ever heard of, photonics.

Speaking of these programs, Robert Reich asserted in an opinion article in the *New York Times*: "Rarely has an Administration sought more actively to encourage specific industries and technologies. Never has an Administration so often justified its interventions by appeals to American competitiveness."[15]

The irony, as Reich well knew, was that such sector-specific policy came in the wake of the political failure and intellectual discredit earned by similar proposals in the brief life of the industrial policy debate. Somehow, sectoral policies directed at specific technologies, including photonics, came into being in several forms for the sake of industrial revival, despite the ostensible rejection of such policy.

PRIVATIZATION AND LIBERAL CAPITALISM

In the 1980s, corporate, academic, and military elites could not help but observe that the relative decline of technological competence (in photonics in the present case) had systemic effects on their institutions' purposes, or at least that this decline could be used as ammunition by which to extract federal resources for responding to the problem. A dramatic range of policies came into effect for responding to the country's declining technological performance, budget deficits notwithstanding. But since such responses had been proved in debate and political prejudice to be illegitimate, they could not take the form of explicit governmental policy. They became politically submerged.

It will be the burden of upcoming chapters to give evidence that privatization took place and to describe how it happened. The purpose here is to elaborate on how we will use the word and to suggest that privatization, if it indeed broadly characterized U.S. economic policy, invites us to revise our thinking about American capitalism. As used here, the word *privatization* encompasses and extends the common usage. It is is usually taken to mean the delegation or contracting of public services to private firms. E. S. Savas, a prominent supporter of privatization, defines the concept as follows: "Privatization is the act of reducing the role of government or increasing the role of the private sector, in an activity or in the ownership of assets."[16] His examples of privatization are drawn largely from city operations and social services: transportation, street services, waste management, education, and so forth.

His general remarks suggest, however, that privatization potentially has broader implications. Since his definition extends privatization to government "activities" in general, not just to the delivery of services, one could extend the meaning even to public policy making. Policy-making, too, is is an activity that could be increasingly delegated to the private sector.

Another author on privatization, Ted Kolderie, argues that such an extension is already implicit in the word. The use of the word *privatization* would be clearer, he writes, if we realized that the term incorporates policy decisions to *provide* a service as well as the administrative actions that *produce* the service. The function of providing the service, as opposed to that of producing it, includes activities of policy making and deciding. He writes: "Since the essence of goverment lies in the first function of deciding what it will provide—what it will require and buy and make available; where and when and to whom and to what standards—*this* is the real (as Butler says, complete) privatization."[17] This real or complete privatization could mean that governmental provision of a service is simply eliminated. The service might then be provided by the market.

But this privatization could also mean the delegation of policy making to private parties, as through a contract. One commentator has referred to such "privatization by contract" as the least clear of the meanings of privatization, a form or privatization liable to be misused.[18]

Through policy-making by private parties, privatization could even yield economic policy that, paradoxically enough, does not turn over an activity to the market but rather places an interventionist policy under private, semiprivate, or public-private auspices. Such privatized technology policy meant for industrial purposes indeed characterizes the public response to photonics.

As the word *privatization* is used in these pages, it also has a further meaning: that of the removal of governmental activities from an open, public realm to a private, enclosed sphere in which an activity that is ideologically suspect can be pursued under a more acceptable guise. The main example of this is industrial policy carried out under the cloak of military preparedness. Such privatization, too, characterizes the response to photonics.

The privatization in the response to photonics necessarily raises a larger question, that of the generality of the observation. Did privatization occur in response to other technological developments and in fields of industrial policy other than technology? One must be circumspect in answering; such are the limitations of a case study. Yet one can, as a preliminary remark, observe that privatization does seem to characterize the U.S. response to other massive, industrially relevant technological developments, like biotechnology and semiconductor technology. And privatized policy making also seems to characterize other types of policies by which the federal and state governments respond to industrial change. These include education and training policy and urban land policy. The concluding chapter of this book will return to this observation. It should suffice for now to state that, from an examination of the response to photonics, one can suggest, not prove, that the general U.S. response to industrial change was similar.

If so, then the privatization of industrial policy might be inherent to the present makeup of the American state. Economic decisions arise not so much from unfettered markets driven by the spirit of competition as from industrial policy operating in forums without vision or accountability. That result runs counter to widely held conceptions of the U.S. political economy in which this country is seen as exemplifying liberal or market-driven capitalism.

Market-Led Liberal Capitalism?

The rejection of industrial policy (but not the sublimation and privatization of such policy) fulfills widely held expectations about the American brand of capitalism. These beliefs have been popularly held as well as carefully delineated in scholarship.

In popular thought, the United States economy operates by reliance on markets. Investment choices arise from private, self-interested decisions. Exceptions occur where government takes a role in such decisions, interfering in the market for overriding public purposes, such as public safety, or because of the vanity of politicians. Government interferes intermittently, and more or less justifiably, in the economy, but the economy remains by a large measure a market economy.

An essentially similar, if more elaborate, picture has been strenuously perpetuated by orthodox economists. But theirs is an elaboration of the conventional picture of the U.S. economy drawn from well-known assumptions about the operation of markets and the distinguishability of public and private spheres. For the sake of contrasting the idea of privatization with other conceptions of the U.S. political economy, a more useful body of scholarship comes not from orthodox economics but from recent comparative work. Unlike the works of professional economists, these comparative studies do not start with the assumption that the market model truly represents the American economy but rather look critically at the political structures through which capitalist nations operate. Their observations reflect a comparative vision that tries to draw out the broad structures of capitalist political economies.

The studies show a willingness to reopen and rearrange the old baggage carried by most observers of capitalism. Their argument that capitalism comes in several models is important for present purposes, because the structures by which capitalism operates come into question. These structures become a subject of inquiry and are not assumed. Their argument is also important because it justifies the present task of characterizing the U.S. response as one kind of capitalist response and not necessarily as prototypical capitalism.

As it turns out, the comparative studies of capitalism reaffirm the usual conceptions of the U.S. political economy. In the works of Peter Katzenstein and John Zysman, this country exemplifies liberal capitalism or market capitalism. To these authors, the U.S. more than other advanced capitalist nations, acts consistently with liberal economic principles ("liberal" used here in the nineteenth-century English and contemporary European sense of unfettered markets). It chooses market solutions to industrial change.

Katzenstein discerns three dominant political forms of advanced capitalism: statist, corporatist, and liberal. Statist countries, such as Japan and France, have a core of elite economic bureaucrats, state-controlled financing for business investments, broad administrative discretion, well-established consultative relations with business leaders, and a range of governmental institutions that concern themselves with economic sectors. All in combination allow the state to undertake policy for purposefully reorienting the economy. The state can take the lead in adjusting specific industries in anticipation of international economic change.

Corporatist countries include most of the smaller Western European states, such as Austria and Switzerland. Dependent on international markets, but unable to develop long-term plans for sectoral change, these countries have to adjust flexibly to international change. They can do so because of their corporatist political structures, which consist most notably of national associations of business and labor. Through ongoing negotiations between these associations and government bureaucrats, in a process that operates in parallel to parliamentary democracy, these countries are able compensate labor for the pains of adjustment and therefore to gain the assent of business and labor to industrial change.[19]

America and the United Kingdom are, by contrast, liberal capitalist nations that respond to global markets through free trade and unhindered foreign investment. "Liberal countries such as the United States," Katzenstein writes, "rely on macroeconomic policies and market solutions."[20] Exceptions are ad hoc and limited. He adds,

> Lacking the means to intervene selectively in the economy, the United States, in those extraordinary situations where the traditional market approach seems to fail, tends to export the costs of change to other countries through the adoption of a variety of limited, ad hoc, protectionist policies.[21]

To John Zysman, too, contemporary capitalism has three predominant forms. He draws his distinctions from an examination of the sources of industrial capital. Typifying his first model, France and Japan have credit-based systems in which the government plays a dominant role in administering prices. State administration manipulates the cost of capital to discriminate in favor of some users and against others as means of implementing sector-specific policies. In his second model, exemplified by West Germany's bank-led system, a few financial institutions dominate financial markets and exercise selective controls.

But Zysman holds that the U.S. (and again the U.K.) seems to conform to the model of a market-led system, where freely traded stocks and bonds are the predominant sources of industrial funds, and where government operates in a sphere separate from the private market.[22]

Writing specifically about technology policies, Henry Ergas also proposes a threefold classification of capitalist states. To Ergas, U.S. technology policy is now classified as "mission oriented," with public science and technology investment being directed at big problems, especially defense, space, and energy. Since the mission orientation concentrates technological investments, research focuses on a small number of technologies, such as aerospace and nuclear energy. At the same time, the diffusion of technology in the U.S., in contrast to that of Germany, Switzerland, or Japan, depends on the market. "In the United States," Ergas writes, "the diffusion of technology is largely a market-driven process, which relies on high levels of mobility of human and financial resources and the existence of a marketplace of ideas."[23]

Hence, in the comparative studies of Katzenstein, Zysman, and Ergas, as in popular opinion, the U.S. exemplifies a liberal or market-driven political economy. It features the separation of public and private sectors into distinct realms. The public realm may intervene in the private one, but it does so intermittently and mainly through generic policies. The rejection of industrial policy, as outlined in the previous chapter, is entirely to be expected and does not invite further inquiry. Not just in conventional economic thinking, but also in a body of comparative thought that has taken a fresh look at advanced capitalist states, the industrial policy debate and the U.S. response to industrial change has occurred in accordance with well-known characterizations of this country as a liberal capitalist nation.

Privatization, Interest-Group Liberalism, and the Contract State

Domestic response to the rise of photonics suggests that the conventional depiction is fundamentally wrong, that this country has an industrial policy, but one that operates in privatized forms. There is, in American political thought, a tradition that recognizes such privatized policy-making. It is found in works by Don Price and H. L. Nieburg on R & D contracting and in the work of Theodore Lowi, particularly in his volume *The End of Liberalism.*

To Lowi, recent U.S. political history is characterized by the abrogation of authoritative policy making through law in favor of policy making led by a host of interest groups. Unlike the nineteenth-century liberalism of the night-watchman state, this interest-group variant of liberalism accepts the positive role of government in economic affairs, as long as government does not operate through the exercise of direct coercive authority. Rather, government makes laws that are purposefully vague, thereby allowing interest groups to play roles in the budgetary process and to penetrate and shape bureaucratic responses to public issues. To Lowi, it is such interest-group policy-making that has produced the massive failures of domestic policy-making of the 1960s and 1970s.[24]

Lowi's interest-group liberalism strongly resembles privatization, as it has been discussed here. As Lowi writes, "Interest-group liberalism seeks to justify power by avoiding law and by parceling out to private parties the power to make public policy."[25] To Lowi, therefore, "partnership," "cooperation," and "participation" of groups in the policy process are code words for the appropriation of policy by those groups.[26] For him, the prototypical example is U.S. agricultural policy, in which public actions emanate from an extraordinarily complex amalgam of committees, cooperative bodies, and interest-dominated bureaus. "Agriculture is the field of American government," Lowi writes, "where the distinction between public and private has become closest to being eliminated." This has been accomplished, he continues, "by private expropriation of public authority."[27]

Commentators on U.S. science policy in the post–World War II years similarly warned that such policy had become the province of shadowy contract relations. To Don Price, "federalism by contract" emerged in U.S. science policy primarily because preparations for intercontinental and atomic warfare had broken the country of its older habit of relying for defense on industrial mobilization after the outbreak of war. Henceforth, the U.S. government would relentlessly develop the technologies of future weapons systems. The method of doing so, that of enlisting companies and universities under contract rather than setting up military laboratories, was originally refined during World War II especially for atomic research.[28]

Furthermore, technological development became so rapid and unpredictable that the field of engineering could no longer survive, as it had in the past, just by mastering a company's traditional requirements. Engineers and the companies that hired them would not just have to make processes work more efficiently but would have to develop new markets. It would be in the companies' and universities' interests to encourage government contracting to obtain R & D funds. This coincidence of governmental and private-sector intentions reinforced the spread of government contracting, contributing to the dissolution of traditional barriers between public and private sectors.[29]

H. L. Nieburg referred more critically to this complex of R & D relationships as the "contract state." Amid national consensus in the 1950s and 1960s in favor of defense technology, space exploration, and scientific research, private contractors penetrated government to ensure continued opportunities and maintain their power. While this R & D contracting complex adhered to shibboleths of free enterprise, it operated through political influence. And while denying the legitimacy of planning, Nieburg wrote, the participants in the contract state did in effect plan, but it was makeshift, backhanded planning scattered among many agencies and serving powerful interests. The outcome was a self-perpetuating power structure, a veritable fourth branch of government that promoted its own interests in the guise of national interest.[30]

According to Lowi, Price, and Nieburg, contracting complexes and interest-group politics have suffused certain American economic sectors—like military and space research and agriculture. These sectors did not operate according to market principles but through a convoluted process that included oligopolistic rivalry, backhanded planning, bargaining, complex social relations among elites, and political pressure.

It seems that the U.S. responded through old formulas of privatization to the challenges posed by advanced technology—particularly our case of photonics—in the 1980s. But unlike agricultural policy, space programs, and weapons development, the privatization of the 1980s occurred in response to the problems of competitiveness encountered in an internationalized economy. It sought to resolve technological problems of U.S. industry without admitting that it was industrial policy. In a world in which other nations take different paths to industrial adjustment, we need all the more to investigate the effectiveness of such policy.

Political Economies and the Industrial Policy Debate

The work of the comparative political economists and Lowi's writings have a prescriptive as well as descriptive inclination. Zysman and Katzenstein provide industrial policy recommendations, while Lowi does so in effect. They take implicit positions in the industrial policy debate. This evaluative intent (evaluation of the policy-making potential of the U.S. political economy) and prescriptive effect (prescription on how the country might conduct industrial policy) pose a central question for the present study: How do we assess the value of and make prescriptions for a privatized political economy?

Zysman holds out explicit industrial policy, in Japanese or French forms, as a potential model for the U.S. With coauthors, he has become an advocate of explicit industrial policy in the U.S., even if he does not always embrace the term.[31] In effect, he returns us to the ill-fated debate over the merits of such policy.

Katzenstein cautiously suggests that the U.S. may profitably borrow elements from the corporatist states of Western Europe. Katzenstein's relationship to the industrial policy question requires more elaboration, since he writes disparagingly of conventional debates in the U.S. over the response to industrial change. He writes that the essential question of the debate has been "Is the state smart or stupid?"[32]

Indeed the industrial policy debate centers on the capability of the state. But Katzenstein's own work raises questions that form the substance of the debate. He writes that his work cannot prove, but accepts as plausible, that corporatist states have successfully responded to industrial transformations because of their ability to facilitate shifts in factors of production. "Flexibility in shifting factors of production," he writes, "presumably strengthens economic performance."[33] But the premise leaves Katzenstein's argument open to the very rebuttals in which the critics of industrial policy have specialized. They would ask why such institutionally negotiated adjustment is more flexible than the sum of the decisions of individual firms. The answer to the critics might lie in the integral character of commonly held productive resources (productive factors) like technology. But the answer would have to be explicitly argued as part of the industrial policy debate.

To Lowi, interest-group politics might be reformed if U.S. policy can overcome its reliance on "distributive policy." Such policies:

> are characterized by the ease with which they can be disaggregated and dispensed unit by small unit, each unit more or less in isolation from other units and from any general rule.[34]

Distributive agencies, such as the Department of Agriculture and the Department of Commerce, serve constituencies through the disaggregation of public resources and can therefore fall under the thrall of such constituencies without deviating from their agencies' formal missions.

To Lowi, the cure for such ineffectual policy-making is a strong dose of stricter, more specific law. Yet it is unclear how the rule of law can respond to sector-specific industrial problems. In some of his work Lowi suggests that legal explicitness can be accomplished through the use of regulatory instead of distributive policy.[35] But if the subject of policy-making is international technological change, then it is not clear that policy-making based on general rules can adequately respond. It is not at all evident that there can be rule-based technology policy for industrial revival.

If regulatory policies are not a practicable substitute for disaggregation, we might be left with distributive policy as the only feasible U.S. response to industrial change. But that result is problematic, since Lowi seems to hold that distributive policy itself is more or less wholly expendable. He holds this view particularly about the U.S. Department of Commerce: a distributive agency operated mainly to respond to the wishes of business constituencies, it amassed the most barren record in U.S. government in the 1970s. Its absence, Lowi asserts, would be felt by no one except, perhaps, its direct beneficiaries.[36]

If regulatory policies cannot adequately respond to industrial change, and if the rest of U.S. economic policy is doomed to disaggregation, then the country simply cannot respond cohesively to the new industrial and technological changes.[37] But what if some classes of common resources—technology, education, land and infrastructure—are essential for the nation's very capability to respond effectively to international economic competition? Then we cannot simply eliminate distributive policy with equanimity but must ask if we can find alternative structures in American policy making. Lowi's work on distributive policy, first published in the 1960s, leaves us with unresolved questions about our capacity for escaping interest-group deadlock to deal coherently with the industrial problems faced in an internationalized economy.

Important work by commentators on the American economy invite us, therefore, to reinvestigate the efficacy or futility of industrial policy, especially if that policy takes place in the U.S. in privatized forms. To do so, we have to confront the industrial policy debate, thereby to examine the principles by which such a valuation can be made.

INDUSTRIAL AND TECHNOLOGY POLICY: PROBLEMS OF DEFINITION

What is Industrial Policy?

Before going on to the industrial policy debate, we should define this much disputed concept and ask what role technology plays in it.

Industrial policy has often and justifiably been criticized for being ill defined. Among the definitions of industrial policy that have been proposed, three of the more important ones have been the de facto, managerial, and sectoral definitions. For reasons introduced below, the sectoral definition will be preferred here, particularly in a formulation that emphasizes the need for a policy-making knowledge about specific sectors. A sectoral definition acutely raises questions about the capabilities of public institutions, the very question around which the industrial policy debate has revolved.

In the de facto conception of industrial policy, "Industrial policy, as we define it, is the sum of a nation's efforts to shape business activity and influence industrial growth."[38] The U.S. government has disparate programs specifically affecting the housing, mining, petroleum, agriculture, railroad, aviation, commercial fishing, timber and maritime industries, among others.[39] Furthermore, countless tariffs, tax exemptions, R & D programs, job training programs, and regulations affect U.S. industry. By refusing to recognize its own de facto industrial policy, the United States manages only to bungle that policy. In the de facto conception, what is needed therefore is not industrial policy per se, since that exists by definition, but a rational and coordinated industrial policy.

Encompassing all policies that affect industry, the definition is flawed. Policies are included even if they are not directed at, but only have the consequence of affecting, a specific industry. Since the definition includes nearly every economic policy, without attributing to industrial policy any characteristic focus of its own, the definition allows for diverse interpretations of what proposals fall under the industrial policy rubric.

Industrial policy in this sense seems to suggest not a delimited set of proposals but the wholesale revision of economic policy-making to make it more rational or coordinated in some sense. It can easily be attacked with the central planning bugaboo. Aaron Wildavsky chooses to respond to this, the vaguest conception of industrial policy. He writes that "there is no such thing as not having an industrial policy. Action and inaction alike affect the condition of industry." Because the concept is empty, Wildavsky says, interested elites use industrial policy to pursue their own ends.[40]

Lester Thurow is one of several prominent supporters of industrial policy who suggest a managerial definition: "Industrial policies are to a nation what strategic planning is to a firm. They outline the basic strategy the nation intends to follow in maximizing economic growth and meeting foreign competition."[41] To Robert Reich, too, "As a theory, industrial policy is closer to the strategic planning model used by many companies than to traditional macro- or microeconomics."[42] Proponents of the managerial view go on to emphasize that they mean flexible response to structural change and foreign competition, and not rigid bureaucratic control.

The managerial language has a reassuring corporate ring that helps dispel the specter of central planning. Thurow goes on, after the cited passage, to suggest tripartite labor-industry-government panels as the means of undertaking the strategy making. The managerial definition, therefore, gives a sense of who would oversee the management of the economy and, to give the flavor of what it might be like, suggests a corporate model. But this definition, too, includes trade policy, antitrust policy, monetary policy, and any other kind of economic policy, whether or not it is directed at a specific sector. It retains the broad sweep of the de facto definition.

The Sectoral Definition

Most of the proposals for industrial policy also suggest that it should be able to focus on specific industrial sectors and adjust the policy response to the sector's current standing and future prospects. Zysman and Tyson make this feature central in their definition. To them, industrial policy ought to be clearly differentiated from aggregate policies that affect all sectors—such as policies meant for macroeconomic stability or for control of the balance of payments. Industrial policy should also be separated from "market promotion" policies designed to improve markets through antitrust, regulation of exchanges, incentives for the mobility of labor, or other means.

According to Zysman and Tyson, "What distinguishes industrial policy is the *government's capacity to evaluate the problem of individual sectors*, not the kinds of policies introduced to solve them."[43] (Stress is in the original.) This quote provides the correct emphasis for present purposes. The distinguishing characteristic of industrial policy is that it operates through the use of knowledge and judgment about the needs, potentials, and effects of specific sectors.

In view of this definition, the multiplicity of de facto programs affecting specific industries may still serve as one of several reasons for coherent sectoral policy. Corporate-style strategic management and tripartite bodies may offer means of carrying out sectoral policy. But the requirement for sector-specific knowledge, not the policy's de facto or managerial status, identifies it as having the distinctive characteristic of industrial policy. Indeed, critics of industrial policy, in search of a reasonably defined sense of what such policy is, have tended to assume the sectoral definition as the most interesting one for their rebuttal.

When it is seen as policy that is purposefully directed at specific sectors, industrial policy becomes a potentially well-defined and manageable proposal. Though pursued de facto in this country, it has not been explicitly reasoned about, researched, or thought through. It can be offered as a supplement to aggregate and market-promotion policies, not as boundless, wholesale reform. As critics have realized, industrial policy as understood by its sectoral definition acutely raises the question of how public policymakers can have knowledge of industrial sectors.

Technology and Industrial Policy

In the debates and writings that appeared in the 1980s on the competitiveness problem, technology policy for industrial purposes was frequently seen as coterminous with and indistinguishable from industrial policy. The conceptual overlapping of technology policy and industrial policy occurred for two reasons. The first was that the industries that provoked the most concern among American policy makers were the technology-intensive industries. The second reason was that the troubles of all industries, not just the elite high-tech ones, came to be diagnosed as technological ills. Analysts differed sharply, however, on what the prescription should be.

Attributions of industrial decline to technology ranged from effusive faith that innovation and better ideas always win the day,[44] to detailed quantitative inquiries on the role of technology in economic growth. Many of the congressional hearings on industrial problems devoted themselves entirely to the role of technology in industrial revival.[45]

The Reagan-era Department of Commerce created an activist Office of Productivity, whose head, D. Bruce Merrifield, spread a gospel of technological innovation and automation. One of many articles and brochures to come out of his office stated: "Innovation is the primary key to increased productivity, and competitiveness in world markets will depend upon continually increasing productivity."[46] As Merrifield wrote in a letter to this author, "The critical understanding . . . is that technical-industrial leadership required for survival in the world marketplace boils down to: Innovate, Automate—or Evaporate."[47]

Reflecting the wide consensus on the importance of technology, Ralph Landau and Nathan Rosenberg wrote in a widely cited collection of pieces that "*Technology* has been *the* critical factor in the long-term economic growth of modern industrial societies."[48] (Stress is in the original.) New technology and a more effective use of technology were seen as essential for the products and production processes of all industries. Furthermore, by definition, the high-technology industries that were of most concern were the very ones most dependent on technological achievement.

Even to economists who blamed the federal budget deficit and high interest rates for holding down industrial investment, interest rates damaged U.S. competitiveness to a great extent by reducing long-term investment in technologies and R & D.[49] Others questioned interest rates as a monocausal explanation, because the slowdown of industrial productivity growth had accompanied emerging trade problems, even in periods when the interest rate was lower. Pointing to productivity as the underlying problem, an influential school of economists frequently saw lagging technology, especially the lagging adoption of process innovations, as essential to a full explanation of the competitiveness problem.[50]

Another explanation had it that American management was the source of industrial decline. Excessive attention to short-term profits and a business-school mentality in which all business resources were interchangeable had, by this explanation, detracted from management's commitment to technological leadership.[51] Technology was even more important to commentators borrowing from the tradition of Joseph Schumpeter. In this tradition, technological revolutions were indeed the driving element in industrial change. Many of the dislocations of the 1980s could be attributed to the transition from production systems dependent on mechanical technologies to new systems of production built on information and communications technologies.[52]

Proponents of industrial policy and its variants relied on several of the preceding arguments. They were also likely to point out that research-intensive industries, (and infant industries working on promising but commercially unproven technologies) were strategic to the success of still other economic sectors. These sectors had to purchase the advanced technologies they depended on for their productivity. In this view, therefore, research-intensive industries were a strategic asset in competitiveness.[53]

Critics of technological capitalism also implicitly acknowledged the importance of technology. To David Noble, much of the technological character of the American economy emerged from military influence over civilian industries in the adoption of technologies of central control.[54] To David Dickson, collusive corporate, military, and academic control over science impeded access by a broader constituency and diverted technology from being applied to social well-being.[55] It followed the tradition of left-leaning critique that Noble and Dickson did not take international industrial rivalry seriously. But, like other commentators on technology's effects on the industrial economy, they saw technology as fundamental.

Technology and industry were seen as being so closely related that Chalmers Johnson, in his definition of industrial policy almost equated the two. Industrial policy at the micro-level sought on the one hand "to identify those technologies that will be needed by industry...and to facilitate their development," and on the other hand "to anticipate those technologies that will decline in importance and to assist in their orderly retreat."[56] To Johnson, industrial policy was defined partly in terms of technology. At the same time, the President's [Reagan's] Commission on Industrial Competitiveness, made technology improvement the first of its recommendations for improved U.S. competitiveness.[57]

Therefore, though the commentators on the role of technology in industry differed in their assumptions and politics, they agreed on the importance of technology. Other than technology, perhaps only education was so widely seen as being essential to the industrial future, and education became a prominent economic issue only late in the decade. Indeed, the very problem of international industrial competitiveness was often defined as a matter of technological rivalry. It seems reasonable to conclude that policies meant to respond to technological change for the sake of industrial competitiveness were a form of industrial policy. (We can refer such policy as "industrial technology policy" for the sake of brevity, but the term should be used advisedly, since "industrial technology" is often taken to mean manufacturing-process technology only.) Such technology policy may well have been the type of U.S. industrial policy most broadly implemented in the 1980s.

Contemporary debates over technology policy for the sake of advantage in trade or competitiveness are descendants from a broader debate at the beginning of the 1980s over industrial policy. There is direct intellectual continuity. And in some ways, the debates in the early and middle years of the decade, before industrial policy became politically discredited, expressed the issues more starkly. Though the present study is directed at the technological dimensions of international competitiveness, the arguments for and against such industrial technology policy are fruitfully examined by placing them in the context of the broader industrial policy debate.

THE INDUSTRIAL POLICY DEBATE

The industrial policy debate dwelt primarily on the question: Is industrial policy worthwhile? There was a flavor of old bromides in the debate. Much of it consisted of making tired distinctions between market and state, market and plan, and (in Katzenstein's terms), stupid state and smart state.

But the distinctions followed the tradition of the most venerable argument for free markets, the argument made by Friedrich Hayek, among others, on the allocative superiority of self-interested actors in unfettered competition. As the argument is commonly cast, it creates a tendentious distinction, contrasting totalitarian central planning with libertarian free markets. Correspondingly, it contrasts the limitations of the state's synoptic knowledge with the flexible, contextual knowledge of firms and entrepreneurs operating in the free market.[58] It is primary among the argument's limitations that it does not distinguish between totalitarian and democratic authority, as if both were vile, though perhaps to differing degrees.

The forceful part of Hayek's argument is not this iron-cast demarcation, but the defense of the market. It holds that freely competing market actors are far superior to government in using information for business decisions. This is a persistent argument, one to be reckoned with. Among criticisms of industrial policy, this was the one to which the supporters of industrial policy had no good answer.

Faced with evidence that industrial policy does in fact occur in the United States in privatized form, persons armed with Hayek's ideas could contend that such policy-making conforms to their worst fears on the incompetence of government. In response to such views, proponents of industrial policy have to do more than say that de facto industrial policy is badly managed.

They have to justify industrial policy by finding a flaw in the Hayekian view of the market. And it has to be a flaw that public institutions can correct with knowledge superior to that available in private firms.

Though they offered many agruments, the supporters of industrial policy did not succeed in convincingly identifying such a flaw. The present section summarizes these arguments. Broadly, there were two kinds. The first consisted of arguments on the evidence. The second kind, which we will consider in more depth here, drew on principled conceptions of the operation of markets.

Arguments on the Evidence

Using evidence from the early 1980s, Lester Thurow argues that U.S. industry is confronting an international economic environment very different from what it faced in the three previous decades. In the face of technologically equal and well-organized competitors, the domestic response suggests "a faint whiff of the stench of economic failure about the American economy."[59] Industrial policy would be warranted, then, by the bare fact that U.S. industry is responding badly.

Industrial policy for dealing with economic failure is justified, so goes another argument on the evidence, because the United States is already doing it de facto. By some accounts, de facto American industrial policy has even been successful in the case of agriculture, aviation, and armaments.[60] More often, American policies of bailing out corporations, creating sector-specific tariffs and quotas, and giving selective federal credits are represented as a failed industrial policy. The prevalence of bad industrial policy is in itself seen as a reason for coherence and coordination to improve it.[61]

Thurow's work illustrates still another argument on evidence in a passage entitled "Others do it—therefore you must do it to remain competitive." Other nations have public investment banking, strategy making through organizations such as Japan's MITI, and government funding for industrial research, so the U.S. must respond in kind. "In a competitive world America has to do what is necessary to be competitive. Competition forces America to do things America may not want to do."[62] Industrial policy is called for because other nations practice it.

The two sides of the U.S. political mainstream responded to the arguments on the evidence in nearly identical terms. Charles L. Schultze, Chairman of the Council of Economic Advisors during the Carter administration, wrote in *The Brookings Review* in 1983 what is probably the most often cited rebuttal.[63] The next year, Martin Feldstein, chairman of Reagan's Council of Economic Advisers, transmitted an *Economic Report of the President* that contained arguments almost identical to those of Schultze.[64]

Both begin by trying to debunk the evidence on problems of U.S. industry. But the international decline of the U.S. technology-intensive industries was just coming to be fully appreciated in 1983 and 1984, when the rebuttals were being published, and became more severe in the years that followed. The stench of economic failure that Thurow spoke of became more intense; the statistical counterarguments were becoming less and less sufficient for clearing the air.

The opponents of industrial policy do not dispute the point that the U.S. has de facto industrial policies. They simply reject the premise that such de facto policies are reasons for explicit policy making. Schultze writes that "One does not have to be a cynic to forecast that the surest way to multiply unwarranted subsidies and protectionist measures is to legitimize their existence under the rubric of industrial policy."[65] Morever Schultze suggests that coordination poses a "false allure." The uncoordinated and ad hoc approach is in fact the correct one, Schultze holds. It is, according to Shultze, a virtue of lack of coordination that each case is treated as an exception.[66]

Both Schultze's article and the Council's economic report also reject the contention that foreign industrial policies, especially MITI, contribute significantly to their countries' industrial ascendance. They agree that other countries have such policies, but hold that there is no proof that the policies actually account for industrial success. Like most debates in the social sciences about causality, this one is irresolvable. Schultze and Feldstein can no more conclusively demonstrate that MITI did not cause the Japanese miracle than Chalmers Johnson or others can prove that it did.[67] Here, the contenders are at loggerheads.

The Arguments on Economic Principle

The more interesting debate was prompted by the second set of industrial policy arguments, which finds flaws in the operation of the market economy.

The first of these argument sees linkages among sectors as a fundamental flaw since, in this view, individual decision makers do not take socially valuable linkages into account in making business decisions. Whether they are semiconductor industries in a world dependent on microelectronics or telecommunications firms in a world built on information, they represent sectors critical to the performance of many other sectors. Moreover, military and civilian sectors are closely interrelated as, for example, in their mutual dependence on advanced microelectronics. Such strategic sectors should be targeted by policy, Stephen Cohen and John Zysman suggest. "The targeted technologies all have one characteristic in common," the authors write. "They are transformative technologies—that is, they are inputs to the products and production processes of other sectors and consequently transform those industries through their evolution."[68]

According to the second of these arguments, markets have another flaw in that they do not adequately incorporate into investment decisions the learning that can be expected to occur during a firm's experience with a technological field or manufacturing process. Early investments in a nation's or region's experience with an industry or technology can give resident firms advantages that pay off when the new industries become commercially successful. Therefore, to create national or regional advantage for the future, policies are needed to identify such sectors early, when private investors are as yet unwilling to invest in such risky fields.[69]

And in the third argument, as expressed by Zysman and Tyson, "The economic rationale for industry-specific policies... is industry-specific market failure." They elaborate that such market failure may take any of the classical forms that economists recognize, namely externalities, public goods, and market imperfections such as oligopolies.[70] Industrial policy is justified in this view because sectors frequently do not operate according to competitive principles.

In their rebuttals, the critics of industrial policy made their counterarguments rest on the issue of knowledge: government is more poorly informed and less able to make good decisions about sectoral characteristics than are private firms. As Schultze wrote, "The first problem for the government carrying out an industrial policy is that we actually know precious little about identifying, before the fact, a 'winning' industrial structure."[71] He adds that "The 'winners' come from a highly decentralized search process, the results of which cannot be identified on the basis of abstract criteria."[72] Though Schultze unfairly characterizes industrial policy with the term "picking winners," a term that the more careful proponents of industrial policy do not subscribe to, he raises the essential point of principle. Sectoral policy by definition requires governmental knowledge of industries or technologies. If government cannot effectively acquire such knowledge, then sectoral policy is a bad idea.

The 1984 *Economic Report of the President* similarly asserts:

> The proposal to target emerging industries suggests that government can obtain better information than the private sector. It is difficult to understand how government officials, together with private business and labor leaders, will be able to gather more accurate information and use the information more wisely than the private sector.[73]

The *Report* goes on to consider some of the separate arguments in favor of governmental influence on sectors. In response to the first argument on linkages among sectors, the report states:

> Some industrial policy proponents advocate government aid to "linkage" industries, that is, manufacturing industries whose output is a vital input into other industries' products. Steel and semiconductors are often cited as examples of "linkage" industries. However, if such an industry is vital, then the industries that rely on it will demand its output.[74]

According to the critics of industrial policy, then, firms and private investors in an industry already incorporate in their actions knowledge of their industry's dependence on other industries for inputs. They also have knowledge of other sectors' dependence on their own output. Governmental action will likely be less well informed about such linkages than private actors are.

The economic advisers offer a similar rebuttal to the argument (the second argument above) on improvements to be gained from learning. The *Report* distorts the industrial policy position by presenting it as a call for protection against imports (a move that most of the prominent proponents of industrial policy disavow). More importantly, it applies to the learning-curve argument the principle of better private knowledge. The *Report* says: "While there may be special cases in which learning curves justify import protection, identifying them in practice would be very difficult."[75] Here again, individual firms and investors are seen as being in a better position than government for identifying learning-curve economies.

Schultze's article and Feldstein's staff preparing the report did not respond to the third argument, the one referring to classical forms of market failure. However, one of the more rigorous critics of industrial policy, Paul Krugman, did respond. And he, too, made knowledge central. In a paper he presented at the Federal Reserve Bank of Kansas City in 1983, Krugman acknowledges that market imperfections would, in theory, justify sectoral policy. But government has no available criteria by which to select sectors for targeting. For each proposed market failure, whether economies of scale, imperfect competition, or externalities, he argues that potential criteria for sectoral targeting do not exist or that the amount of study needed for identifying such sectors would be excessive. "To act with any hope of success," writes Krugman, "would require a deep study of each industry in question—a deeper study than any which has ever been carried out."[76]

In a later volume edited by Krugman on the newly emerging field of strategic trade policy, one of the intellectual progeny of industrial policy, the economic analysts among the contributors are more open minded about such strategic policy, but they finally reject it, each time with an argument about the capacity of public institutions to compile and use knowledge. Gene M. Grossman writes, for example, "Treating the case for strategic export promotion on its own terms, I will argue that we do not now (and may never) have sufficient knowledge and information to merit the implementation of a policy of industrial targeting."[77]

Clearly this, the assertion that public knowledge is not up to the task of formulating industrial policy, is fundamental. It is the rebuttal to which the industrial policy advocates have no good rejoinder.

Answering the critics who point out the absence of good criteria for targeting sectors, Thurow agrees that there can be no criterion that is universally valid for selecting sectors. But "they [the opponents of industrial policy] are wrong in that consistent, rigidly held criteria as to what industrial policy should do are not what industrial policies are about."[78] Holding a managerial conception of industrial policy, Thurow is quite right to reject a call for criteria. After all, corporate managers select investments, R & D projects, and interindustry relationships without having to document their choices with preestablished criteria. Yet, even if industrial policymakers make their decisions through the exercise of strategic judgment, and not through criteria, the fundamental criticism retains its force. Why should policymakers be able to make better judgments than entrepreneurs and inventors?

For supporters of industrial policy, the expedient response is to reject the assumptions of orthodox economics. In responding to Krugman's 1983 paper on industrial policy, Robert Kuttner, journalist and economists' gadfly, does just so: "If we assume, ex-hypothesis, that markets are the optimal allocators of capital, that economies are at or near full employment, and that other nations do not rely on mercantilist approaches, then there is no need for industrial policy."[79] To Kuttner, a world of interbred politics and industry puts the economists' simplifications out of the realm of policy prescription.

A rejection of economic assumptions of course does the job. But for a supporter of industrial policy, a more satisfying rejoinder would be one that found a specific shortcoming in the orthodox economic reasoning. This would have to be a critical shortcoming in the argument on the private actor's superior technical knowledge.

Government's Incapacity, the Clincher?

The critics of industrial policy also had a clincher, or what seemed like one: the idea that government generally, or a U.S. government well known to be in the thrall of interest groups, was inherently incapable of anything that would require action as systematic as that entailed by a coherent industrial policy. And while the economic rebuttals on evidence and economic principle generated some focused debate, this, the political rebuttal, met with remarkably widespread consensus. Government would be incapable of the coherent policy-making needed to carry out industrial policy.

According to Schultze, "Not only would it be impossible for the government to pick a winning industrial combination in advance, but the attempt to do so would almost surely inflict much harm." To Schultze, the reason it would do so is that democratic political systems, especially the U.S. system, do not do well in making explicit policy choices that help some interests while penalizing others. "Yet," Schultze writes, "such choices are precisely the ones that would have to be made—and made explicitly, for an industrial policy to be more than a political pork barrel."[80]

Such warnings of government's inability appear again and again in criticisms of industrial policy.[81] Particularly withering criticism emanates from the U.S. school of political pluralism. *Pluralism* may be defined as the idea that unhindered interest-group access in a democratic political system creates a self-regulating process of political representation; this representation produces public decisions more intelligent than would be made through any systematic planning. But the pluralist school's representatives hold out no hope for pluralist industrial policy. The unhindered contest of interests may be good at budget allocation in traditional fields of government policy, but not in industrial policy, which impinges on decisions traditionally left to the market. Self-regulation in the market is apparently superior to self-regulation in the polity.

Spokesmen for pluralism, like Aaron Wildavsky and Eugene Bardach, treat industrial policy with scorn. To Wildavsky, for example, industrial policy proposals have so little merit that he views them only as a manifestation of American political culture, not as a policy worthy of consideration on its merits.[82]

If industrial policy inevitably degenerates into pork-barrel politics, and if strict regulation cannot replace it, then even a prominent critic of pluralism, Theodore Lowi, would have to agree that industrial policy would become politically bankrupt. Finally, some of the supporters of industrial policy also accept that industrial policy inevitably stumbles over politics. "I agree with the critics," Thurow writes, "that this is the real objection," (but he goes on to refer to this view of politics as "defeatism of the rankest sort").[83]

The argument on the fragmentation of the American polity surely hurts the industrial policy proposal. Moreover, if one first states the principled economic conclusion that the state has inherently poorer economic knowledge, then the political rebuttal is hardly even needed. It is beating a horse that's already dead.

But what if industrial policy making in the U.S. did not arise in a rush of diverse interests but occurred through privatization of policy-making? It may be disturbing as a view of American politics, but it is not readily dismissed as inefficacious in policy-making. And what if, in response to integral resources like photonic technology, public bodies could exercise synoptic knowledge that private firms cannot efficiently make use of? If so, the horse could rear its head again. The proper question would then be how to cure it rather than how to bury it.

THE CONCEPTION OF DISCRETE INNOVATIONS

In critical comments about industrial policy, Mancur Olson raises the principled objection to industrial policy in pure form: "It is primarily in the areas of uncertainty like high technology and new industries that private venture capital has the greatest advantages." If proposals for industrial technology policy are implemented, such policies would give interest groups an opportunity to exploit government largesse. Therefore, Olson demands, "The authors of these proposals should also explain why a government board or agency would allocate capital more effectively than those people and firms that are investing their own money."[84]

In Olson's conception, investors and inventors are closer to the enormous complexity of technical feasibility and market demand than any government policy makers would be. After all, each innovator works on discrete technical projects, each project enmeshed in a unique technical and economic context. National economies then become more successful to the extent that they let innovators (firms, inventors, entrepreneurs) develop the technologies they see as promising. Policies to suggest otherwise will have to demonstrate a source of knowledge better than individual innovators have.

Olson, like other critics of industrial policy, draws his most fundamental objection from a certainty that private economic actors are superior to public entities in making economic decisions, including technological ones. But there is an unspoken assumption concealed in that certainty: it is that the knowledge by which firms operate, and technological knowledge in particular, is multifarious, fragmented, discrete, and context specific. The assumption is that technology is an aggregate of discrete innovations.

This conception of technology should be contrasted with the one presented in chapter 3, where photonic technology is depicted as having integral properties. The argument in chapter 3 intended only to lend plausibility to this conception of integrality; it cannot be said to have provided proof. However, the more common idea of technology as an aggregate of discrete innovations, one that is widely held and inexplicit, is no more proven.

The rest of this chapter articulates the implicit conception of discrete innovations held by critics of industrial policy. It presents the assumptions of technological discreteness as seen in the work of microeconomists and in the more elaborate views of those who seek to describe what is inside the "black box" of innovation. The chapter then concludes with the suggestion that good reasons for industrial technology policy can be found in those dimensions overlooked by the critics: the integral characteristics of technology.

Technological Discreteness in Microeconomics

Conceptions of atomistic innovation are particularly seen in the spare assumptions that microeconomists make about technical progress at the level of the firm. At this level, technology consists of the knowledge through which inputs are transformed into outputs. Through R & D, firms increase the efficiency with which they transform inputs into outputs, thus becoming more productive. Since firms can garner profits by becoming more productive or entering into new product lines for which there is demand, they have an incentive to engage in technological innovation.

The production of such innovative technology is an economic outcome that, like other kinds of production, reflects the convergence of demand and supply. Demand may increase through the rising cost of an input, causing the firm to seek processes that make less use of the input. The supply of means of solving the problem includes scientists, engineers, and a stock of applicable technical knowledge. Technological innovation occurs through the interplay of such demand and supply.

The accumulation of acts of innovation by firms and entrepreneurs yields technical progress in the economy. At any time, society's pool of technological knowledge is the sum of accumulated inventions. Hence, just as the sum of technological innovations at one time can be seen as a pool, technical progress can be seen as a flow or rate.[85] The molecules in that pool or flow are acts of innovation.

Firms not only create technical improvements through internal R & D, but also through the purchase of pieces of intellectual property. Some indeed specialize in producing components and intellectual property to be used by still other firms. Their outputs, such as lasers or microchips, are discrete inputs in other industries; these industries themselves provide inputs, such as office equipment, that the laser and semiconductor firms themselves use. These interindustry flows can be understood as accumulations of discrete events.

For a firm, the availability of such technological inputs and the demand for its own technological outputs are part of the total context of information within which optimal allocation decisions are made. It follows that, in a well-operating market, firms are the institutions best positioned to optimize investments in new technologies, no matter the input-output relations with other firms. The existence of interindustry technological flows are of no unusual economic significance. They do not entail public support for industries having an especially large number of linkages with other industries.

The commonly admitted exception to superior private efficiency in innovation is the case of public goods. Technology is a public good because imitators of the technology cannot be excluded. If patent protection is not sufficient, one firm can take advantage of others' investment in innovation by copying it, without having to make a new innovative investment. Knowing that it will not reap all the economic benefits of innovation, a firm will invest less in innovation than is socially optimal. Public intervention is warranted in the form of a subsidy to encourage private companies to perform more R & D, but the content of that R & D is still best determined by individual companies themselves.

Adherents to such views readily admit that there are distortions in this picture, since government influences R & D decisions and academic researchers affect technology without necessarily responding to purely economic motivations. Yet, the picture is seen as the fundamental one, to which governmental and university research must be introduced as distortions.

Beyond the Black Box

The foregoing understanding of innovation has come under criticism, even within the world of economics, for taking the actual process of innovation for granted. Nathan Rosenberg has been prominent in this criticism. With Stephen Kline he writes,

> Economists have, by and large, analyzed technological innovation as a "black box"—a system containing unknown components and processes... They have devoted very little attention to what actually goes on inside the black box; they have largely neglected the highly complex processes through which certain inputs are transformed into certain outputs (in this case, new technologies).[86]

Kline and Rosenberg propose a more realistic description, one that more truly reflects the complexity of the process, since "The systems used in innovation processes are among the most complex known (both technically and socially), and the requirements for successful innovation vary greatly from case to case."[87] This more realistic depiction should show how innovation necessarily combines technical and economic considerations and uses no single formula but a complex of varying ideas and solutions. According to Kline and Rosenberg:

> The whole process of technical innovation has to be conceived of as an ongoing search activity—a search for products possessing new or superior combinations of performance characteristics, or for new methods of manufacturing existing products. But this search activity is shaped and structured in fundamental ways not only by economic forces that reflect cost considerations and current supplies of resources, but also by the present state of technological knowledge, and by consumer demand for different categories of products and services.[88]

It is this view of the complexity and uncertainty surrounding innovation that corresponds well with the views of those leading researchers and industrialists who are often in a position to affect U.S. technology policy. It is indeed such views of complexity and situated knowledge that prompt calls for public support for research, but no public interference.[89]

The attempt to open up the "black box" by describing its operations as a complex search process even reinforces the prevailing conception of atomistic research. Individual researchers, research managers, entrepreneurs, and venture capitalists are closest to this enormously complex decision-making process and should not be second-guessed by those governmental policy makers operating from a greater distance. Therefore, the attempt to throw light on the process of technological innovation may have made the boxes less opaque, but has left them as piles of discrete boxes nevertheless.

Anticipating Technological Change?

Whether in the spare assumptions of microeconomists or the intuitively more realistic description that stresses the complexity of technological investment decisions, critics of industrial technology policy assume a world of discrete technologies that have economic effects in the aggregate. Even in the more sophisticated conception of technology found in the work of Kline and Rosenberg, technological change arises from the actions of individual innovators acting between the twin constraints of technological feasibility and market opportunity. Innovation in this sense reflects judgment made amid multiple uncertainties. The individual innovator and the firm are the ones close enough to this complex context to make the best judgments on R & D projects and product development investments.

So it is all the more curious that Rosenberg refers elsewhere to predictability in technological development. He makes this point, interestingly enough, about laser and optical technology. He writes that "the advent of the laser as a potentially new mode of transmission has served as a powerful focusing device in shaping the direction of scientific research." Numerous unanticipated applications arose from this new technology, he adds. (Apparently unaware of the extent of early military involvement in laser research, he lists defense and space applications among unanticipated uses.) "At the same time," he continues,

> a high-payoff application of laser technology *was* clearly anticipated and successfully consummated. It was the development of laser technology that suggested the feasibility of using optical fibers for transmission purposes. This possibility, in turn, pointed to the field of optics, where advances in scientific knowledge can now be expected to have high potential payoffs. As a result, optics as a field of scientific research has experienced a great resurgence in recent years. It has been converted by changed expectations, based on past and prospective technological innovations, from a relatively quiet intellectual backwater to a burgeoning field of scientific research.[90] (Stress is in the original.)

A similar but more commonly stated view of the orderliness of technological development is found in the idea that computer technology progresses in generations—we are now in about the fourth generation. But how can the sum of individual decisions under extreme uncertainty take on the character of predictability? Developments in advanced optics and the successive generations of computers can be anticipated if there are structured relationships among innovations, if technological innovations have cohesive and integral characteristics. Hence, the very possibility of anticipating trends in photonic technology suggests that innovations take place within broader technological paradigms.

As argued earlier, it is this technological integrality that should be central to our understanding of massive technological changes, such as the rise of photonics. Photonics gains such integral characteristics as it comes into use through the elaboration of integral bodies of knowledge and practice. Photonics evolves through the development of standards of practice, rules of thumb, theoretical insight, a cadre of educated engineers, appropriate instrumentation, professional accreditation, and research institutions. As elements of the paradigm come into being, the directions of progress in the technology become clearer.

Photonics also comes to be applied through technical systems, such as telecommunications and computing systems, whose development requires standards, equipment, and infrastructure investments that allow interconnectivity. When components of such a technical system are developed, the forms of the rest of the system gain a measure of predictability.

Seen as bodies of knowledge and skill, and as technical systems, photonics is no longer adequately understood as the output of any single industry or group of industries. It is more fruitfully seen as a paradigm that affects the fortunes of numerous branches of the economy. And once the integral logic of the paradigm becomes comprehensible, we can, to an extent, anticipate the range of its industrial effects.

A state that can build the capability for anticipating such technological change will find itself at an industrial advantage relative to other states. If this depiction of photonics is right, then technologies, seen as technological paradigms, have a property quite unexpected by the theoreticians of aggregate innovations: The property of reduced uncertainty. Giovanni Dosi notes this characteristic of technological paradigms as follows:

> When a technological paradigm is established, it brings with it a reduction of uncertainty, in the sense that it focuses the directions of search and forms the grounds for formatting technological and market expectations more surely.[91]

If this relative predictability comes in the form of massive technological developments affecting multiple industries in nations and regions, then those areas potentially have the intellectual means by which to anticipate, and hence plan for, technological change within a capitalist economy. Through purposeful industrial policy, regions and nations can create the institutional capability to scan technological developments in the world, formulate visions of technological trends, and insert a dose of planning into technological investments. The argument for the integral characteristics of technology can, then, answer each of the three criticisms of industrial policy that derive from the principle of the private economic actor's superior knowledge.

The first argument lodged by the critics of industrial policy is that interindustry linkages are best recognized by private firms. The proper rejoinder is that such interindustry effects do not just take the form of discrete linkages but also of structural transformations that affect multiple industries. In the second argument, the critics say that individual firms are most adept at recognizing long-run learning-curve economies. The appropriate rejoinder is that the learning inherent in massive technological change occurs through the development of technological paradigms. These come into being through the operations of nonmarket, as well as market, institutions and evolve through an integral logic to which private firms respond better when they do so in conjunction with a broad public effort to anticipate technological changes.

In the final argument, the critics doubt the ability of government to pick out promising sectors. Even in the presence of market failure, private actors are seen as being much better at making investment choices. The proper rejoinder here is that the opportunities and needs of sectors often reflect the transformative effects of technologies having an integral character. If a nation or region is to respond appropriately to such changes, it should undertake broad analysis, tracking, scanning, and planning—the very synoptic activities that are best carried out by public-purpose institutions.

The concept of technological paradigms (and potentially a concept of other comon resources) offers the seeds of a comprehensive answer to Mancur Olson's challenge. Public policymakers can potentially allocate money more effectively than self-interested actors when responding to massive technological changes characterized by integrality.

Clearly, then, a political economy so structured as to excessively depend on innovators and firms to guide technological development would be less effective in taking advantage of technological paradigms. Industrial policy is called for. But what about a political economy where industrial policy-making takes place but is privatized?

QUESTIONS OF PRIVITIZATION AND POLICY-MAKING KNOWLEDGE

The rise of photonics occurred during the 1980s, a decade of declining U.S. performance in several industries and technologies and growing domestic concern about American industrial standing in the world. Yet this concern could not be translated into explicit industrial policy because of the widespread notion that such a policy would be ineffectual or ideologically illegitimate. Caught between the opposed forces of a need for a public response and the perceived illegitimacy of that response, governmental and industrial elites responded to photonics with privatization.

In privatized industrial policy, policy-making shifts from a public and open sphere to a private and closed one. The privatization of industrial policy, then, does not return resources to the market, but, ironically, places an interventionist economic policy under collaborative, public-private, and military control.

This idea of privatization draws on a tradition of political analysis that has portrayed the American political economy in an unusual light. In this picture, complexes of contractual relationships dominate important U.S. economic sectors and private interests penetrate and appropriate public policy-making. This model does not intend to depict capitalist political economy overall, only the particular institutional makeup of American capitalism.

Even though recent comparative studies of industrial nations have described the United States according to a more stereotypical market model, these studies have made the enormously important observation that capitalism is not just one kind of economic system. It has varied forms; the privatized arrangements of the American political economy can be considered one of these forms. Not the least reason for the importance of the observation is that the structures of capitalism have policy implications. Such models of American capitalism suggest whether and how the country can use public policy to respond to industrial problems. Having a prescriptive dimension, these models revive contentious issues in the industrial policy debate. It is this debate that most starkly asks whether industrial policy in the U.S. would be worthwhile and workable.

To examine the debate, we need a definition of industrial policy. The following definition turns out to be the most useful: Industrial policy is that set of economic policies that requires government to evaluate the problems, potentials, and effects of specific sectors. The definition raises the question of government's capacity to make such sectoral judgment effectively.

The question is crucial, since the principled arguments of the industrial policy debate revolve around it. Advocates of industrial policy suggest that interindustry linkages, the capability for learning in the economy, and sector-specific market failure all justify industrial policy. Opponents retort that firms, entrepreneurs, and investors are best positioned to recognize interindustry relationships and learning-curve economies. And, they argue, even where sector-specific market failure occurs, government has poorer sources of information than do private, self-interested actors.

This argument, when unpeeled, contains an ideological kernel. It conceives of technology as the aggregate of discrete acts of innovation occurring in highly complex, unique situations. Yet, photonics can be characterized quite differently: not as an aggregate of discrete innovations, but as a technological paradigm with integral properties. It is a further characteristic of a technological paradigm that, within it, technological innovations become less uncertain. When nations can respond with public policies to the changes brought on by the technological paradigm, then their industries can gain advantages with respect to rivals who do not benefit from such policy. If this conception of technological paradigms is reasonable, then we have an economic argument for industrial policy.

The argument is essential if we are not only to observe that industrial policy became privatized in the U.S., but also to pass judgment on it. Without the argument on technological paradigms, privatized industrial policy might just seem to be the fumbled political outcome of a policy that was a bad idea from the start. With the argument, however, we can both observe the political economy of privatization and evaluate the programs it fosters.

The argument allows us to pose critical questions for assessing how U.S. policies have responded to photonics. First, do U.S. responses really constitute an industrial policy, if industrial policy is given the sectoral definition outlined above? Second, is such policy really carried out in privatized forms, and what indeed are these forms?

Sector-specific knowledge is the criterion by which to evaluate industrial policy. This leads to the third question, how do privatized forms of policy making acquire the sector-specific knowledge by which they carry out industrial policy? And, finally, can such privatized policy-making fulfill the essential task of sectoral policy in capitalism: Can it respond coherently and with vision to the integrality of massive technological change? The chapters that follow respond to these questions. They are arranged in roughly ascending order, from disaggregation through collaboration to sheltering, from lesser to somewhat greater ability to knowledgeably respond to photonics and its industrial effects.

FIVE

Disaggregation

DISAGGREGATION, SECTOR-SPECIFIC AND OTHERWISE

Among forms of privatization, "disaggregation" refers to the most traditional and politically expeditious kind of response to public issues. In disaggregative policy, governmental assets are dispensed to interested parties unit by unit, each unit in isolation from the others and from any encompassing strategy or rule.[1]

When the issue at hand is domestic technological standing in foreign trade, the disaggregative response is to spread resources to technologically inclined private companies, university research programs, and individual researchers. The disaggregation may occur in kind or in cash. In-kind disaggregation makes government research laboratories and government-owned intellectual property available to interested parties. Disaggregation in cash, a plain enough concept, provides tax incentives, grants to specific institutions, and grants distributed through review committees consisting of anonymous representatives of the interested parties.

These are well-tried and well-known policies, so their intellectual underpinnings are not always strong. When intellectual support is expressed for such policies, it operates on the assumption that technological innovation is cumulative, incremental, and unpredictable. By this idea, technical progress arises in aggregate from the R & D decisions of disparate scientists, engineers, and businessmen. Businesses and their technical specialists, not government bureaucrats, are best qualified to judge technical needs. Policies of governmental support for R & D can therefore distribute funds back to those who do R & D or profit from it, letting them decide the research agenda. Since the thinking about the purposes of technology is shifted to private interests, the disaggregative form of privatization is characterized by minimal amounts of coherent thinking among policy makers about where research resources go or what purposes they serve.

Amid the rising tide of trade problems in the 1980s, when national and regional technological strength came to be seen as a prime means of regaining industrial ascendance, disaggregative policies came to be applied in ever greater breadth and variety to private research-performing organizations. The policies included tax incentives for research, transfer of government technology to private firms, pork-barrel funding of university research centers, the opening of government laboratories to private firms, and allocation of research funds through review committees. Since photonics grew in importance during the same time, there were many instances in which disaggregative policies were used for expanding research in photonics.

In two of the clearest examples of disaggregative policy, the tax incentive and transfer of governmental intellectual property to private interests, governmental assets are distributed without intended prejudice to specific economic sectors. Since such policy clearly transfers government assets to the market and pays no heed to sector-specific needs, it is not industrial policy. The rest of this section briefly reviews these two purer types of disaggregative policy. The sections that follow then describe disaggregative policies that are indeed examples of industrial policy. Somewhere, somehow, such disaggregation of governmental assets incorporates decision-making about specific sectors, including photonics.

Tax Incentives and Technology Transfer

Technology policy as the disaggregation of government assets has been exemplified in purest form by the R & D tax incentive. As legislated in the Economic Recovery Tax Act of 1981, taxpayers were allowed to claim a 25 percent credit on the increases in certain research and development expenditures. In the Tax Reform Act of 1986, the credit was reduced to 20 percent, and qualifications for it were made more stringent, but a separate credit was added for corporate research grants to universities.[2]

As estimated by federal budget documents, the R & D tax credit in 1988 added up to the equivalent of a federal expenditure of $1.06 billion.[3] Being purely disaggregative, that expenditure reflected no special consideration of any technical research area. If it affected the practical affairs of research performers, it did so only by exposing them to the rigors of the Internal Revenue Service's Form 6765. In keeping with the idea that private firms were best placed to judge research needs, even if their researches frequently reflected military rather than commercial priorities, the policy took no position on photonics or any other field.[4]

Another set of policies, which were implemented widely for the first time in the 1980s, would distribute federally owned intellectual property to private organizations. Like most of the technology policies that emerged in the decade, this one, too, did not reflect the priorities of only the Republican administration. It had its legal inception under Jimmy Carter. Late in his administration, the Patent and Trademark Act was amended to allow nonprofit organizations (mainly universities) and small businesses to retain patents issued for federally funded research.[5]

A presidential memorandum issued in February 1983 directed federal agencies to do all permitted by law to enable federal contractors and grantees to retain title rights to inventions. Further amendments promulgated in 1984 broadened the waiver of exclusive federal ownership by allowing most of the Department of Energy's giant laboratories, operated by nonprofit organizations, to also retain rights to patents obtained under federally sponsored research.[6] The federal Technology Transfer Act of 1986 authorized government-operated laboratories to cooperate on R & D with other organizations, including businesses, and grant these partners title or licensing rights to inventions.[7]

The Technology Transfer Act also formally chartered the Federal Laboratory Consortium for Technology Transfer, a loose organization of federal officials interested in technology transfer issues. They were to act as a clearinghouse to direct private inquiries to federal laboratories performing research in the appropriate area. For example, a caller in 1989 interested in federal labs pursuing optoelectronics research would have been directed to a consortium member at the U.S. Air Force Photonics Center at Griffiss Air Force Base in Rome, New York.

The National Technical Information Service, a branch of the U.S. Department of Commerce, handled licensing arrangements for most federal civilian agencies' patents. Even this agency, the most important government licenser, granted only between twenty and seventy licenses each year between 1981 and 1991. For patents that the organization licensed, it made decisions on whether a license was to be given exclusively or nonexclusively to firms that applied. If no one objected to a notice placed in the Federal Register, an exclusive license could be granted to an applicant company.[8] Unlike large industrial corporations, which made patent disclosures and sought clusters of patents to protect interests in selected technical fields, the government policy treated patents as discrete items having no relationship to each other or to the needs of specific sectors.

The Sections Ahead: Disaggregative Industrial Policy

In both tax incentives and transfers of federal intellectual property, but more in the former than the latter, disaggregation occurs in pure form. With little or no consideration or selection government assets are transferred to private claimants. One can barely identify whether the policies impinge on specific technological developments, such as optoelectronics. Such policies are not industrial technology policy.

In other kinds of disaggregative policy, one can identify some political or peer-review process that takes photonic technology into consideration. Whether through political moves to earmark budget allocations for academic research facilities, or through business participation in the work of government labs, someone, somewhere makes decisions that emphasize one technical field rather than others.

The sections that follow examine sector-specific disaggregation. They ask how decisions are made to direct governmental assets toward photonics and (where relevant) how the choice of research projects relates to the purpose of industrial competitiveness. The sections examine legislative earmarking in photonics; arrangements for business participation in public research; and academic research funding through "merit review," the recent term for peer review.

OPTICS AND PORK

Pork Barrel Resurgent

The most elusive of government policies are those hidden away on wayward budget items, appropriations bills, and associated committee reports. When they are used to fund physical facilities in a legislator's home district, such budget items are traditionally referred to as "pork-barrel" or, more politely, as "earmarked" appropriations. Pork barrel or earmarking may be defined as budgetary appropriations, usually for physical facilities, that specifically serve a regional constituency and that pass through no review that would fit them into a broader purpose.

As technological competitiveness arose as an urgent issue in the 1980s, so did a new form of pork barrel that responded to perceptions of technology as a linchpin of local prosperity. Referred to by one journalist as "postindustrial pork," earmarked federal funding for university facilities, especially science facilities, leapt from $2 million in 1982 to $291 million in 1988, budget deficits notwithstanding.[9] In 1989, earmarked federal funds amounting to $289 million went to ninety academic institutions.[10]

The hoary tradition of the pork barrel represented one response to the rise of photonics. The cases were few, but each was substantial. State and federal governments made discrete appropriations to specifically named institutions for buildings and equipment for university research in optics, imaging, and optoelectronics. In this kind of policy, decisions to pursue photonics research were the outcome of the stratagems of the academic institutions themselves and of legislative negotiations in which specialized lobbyists had a critical role.

Earmarking for academic institutions had long existed, but it had formerly been reserved for colleges having traditional relationships with the federal government, such as Gallaudet College, Howard University, and Tuskegee Institute.[11] Until the early 1970s, the federal government funded campus construction programs through peer-review procedures carried out by several agencies, primarily the National Science Foundation, but these programs were discontinued under the Nixon administration in favor of aid to individual students.[12] There have since been no systematic sources of federal funds for college facilities (though a new federal initiative, authorized but not funded in 1988, may yet reinstate some facilities funding through peer review).

Over the same period, the late 1970s through the 1980s, technology came to be seen as much more important to regional prosperity. As stated by the president of the Association of American Universities, a group opposed to pork-barrel funding, "The fat is in the fire. As the perceptions of a connection between science and technology and economic development intensified, science has been transformed from a private activity to one useful for achieving universal goals."[13]

Earmarking for Optics: RIT and Huntsville

Many of the earmarked projects were devoted to specific technical fields. Of earmarked appropriations between 1980 and 1989, two can be identified as being clearly devoted to photonics. Public Law 99–190, Making Further Continuing Appropriations for FY 1986 and for Other Purposes, specified $11.1 million for "microelectronic engineering and imaging sciences at Rochester Institute of Technology," (RIT) with no further comment. Public Law

100–202, Continuing Appropriations, FY 1988, referenced $10.6 million for the Center for Applied Optics, University of Alabama.[14]

On the RIT campus in Rochester's southern suburbs, the fruits of congressional earmarking took shape in 1989 in the form of a new red-brick building. It houses the nation's first undergraduate program in imaging and the Institute's very first Ph.D. program, also in imaging.

RIT's decision to go forward with imaging emanates from developments in its nationally respected School of Photographic Arts and Sciences, which has long taught the art and technology of photography. Through the 1970s, the school's photoscience department emphasized technologies of photographic emulsions and darkroom work. In the late 1970s and early 1980s, members of the faculty began to see that their work on electronic imaging technologies had an array of applications in remote sensing, medicine, and industry.[15]

As intradepartmental disagreements arose over the technical directions of the photography program, M. Richard Rose, former head of Alfred University and former assistant secretary of defense for education, established himself as the new president of the institute.[16] Rose decided to champion the photography faculty's new technical interest in order to make imaging technology part of the institute's effort to build a specialized technological niche through which RIT would gain national stature.[17]

Rose's administration chose to build its technical standing in imaging (and in another field, microelectronics) through directly earmarked federal appropriations. As do many prospective applicants for earmarked funds, RIT reached the congressional ear through the services of one particular lobbying firm, Cassidy and Associates, a company that has built a reputation in helping colleges obtain earmarked funds and maintains a long client list of academic institutions. Once a college becomes a Cassidy client, states Roy Meyers, spokesman for the company, "we help them develop a proposal or prospectus and work with them on bringing it to the attention of congressmen, especially local representatives."[18] The company does so by retaining as consultants several former college presidents and former congressional staffers. Having obtained a legislator's interest, says Meyers, "we attempt to form a coalition."

The lobbying effort brought New York's Republican Senator Alfonse D'Amato's assistance for RIT's cause. He had an $11 million earmark inserted in a $12 billion appropriations package in 1985. This, however, required reciprocal agreements with other legislators. Negotiations over the package brought together Paul Laxalt, Republican of Nevada, pressing for an engineering facility at the University of Nevada; Edward Kennedy, Democrat of Massachusetts, supporting a project at Northeastern University in Boston; and Ernest Hollings, Democrat of South Carolina, who wanted a fiber optics center at the University of South Carolina, though Hollings's initiative appears to have stalled.[19]

After a congressional appropriations committee earmarks funds for specific facilities, one of the agencies under its jurisdiction is made responsible for disbursing the funds. The Department of Energy's Office of Energy Research is a popular vehicle for earmarked appropriations. After each appropriations cycle, the office receives a list of facilities it must fund and has no further say in the matter, except that it requires the academic institution to submit a construction proposal that conforms to federal regulations. This office handled the funds for the University of Alabama's Center for Applied Optics.[20] The Department of Defense administered the money for RIT's Center for Imaging Science.

Science Earmarking in New York State

State governments have their own traditions of science pork. In New York State alone, between 1984 and 1989, discrete appropriations for university high-technology facilities exceeded $200 million (in grants and low-interest loans). Unlike federal earmarking, which at least in the early part of the decade favored the less well known campuses, New York State's earmarking showed no particular orientation for or against campuses that were favored by conventional grant sources.

Earmarked grants were given to the State University of New York at Buffalo for an earthquake research center, Cornell University for a supercomputer building, and Rensselaer Polytechnic Institute for an industrial innovation complex.[21] In 1987, a long list of discrete facilities projects appended to the state's Omnibus Economic Development Act included a $10 million appropriation for the University of Rochester's Center for Electro-Optic Imaging, the purpose or function of which was not further described in the act. Construction of a building by that name was set to begin on the university's campus by 1990.

Each institution that wanted to obtain state funds for science buildings had to pursue earmarked grants, since New York provided no formal mechanism for the applicants; lobbying was the only means available. To obtain the state grant for its electro-optic facility, the University of Rochester had to "hit all the bases," reports a high ranking budget official who was close to the process. Working with Eastman Kodak's Albany lobbyist, the university approached local representatives, Senate and Assembly leaders, and contacts in the executive office. And the timing was right. State lawmakers in 1987 had surplus funds and sentiments favorable to high-technology issues. The few universities that appeared with proposals for new facilities that year were funded. The state's Urban Development Corporation was given the task of evaluating proposals—after the appropriation—and working with the institutions to prepare financing plans. "As for technical work to decide on proposals, none was done," recalls the budget official.

News of the success spread. By 1988, "the process got out of hand. Everyone and his brother started coming forward wanting money."[22] Among the claimants was a delegation from Rome, New York. Having undergone a painful industrial restructuring and the loss of manufacturing enterprises, Rome hoped to partake of high-tech growth. But the local unit of the state university enjoyed no fame. Rome, therefore, hoped to build on the Air Force's designation of the Rome Air Development Center, at Griffiss Air Force Base, as the Photonics Center for the entire service. Local representatives lobbied hard in Albany for $18 million to build a photonics building that would spin off Air Force innovations into local growth.[23]

During spring 1988 a flood of new proposals, such as the one from Rome—adding up to more than $100 million—were submitted, dismaying the state's most powerful political leaders, Assembly Speaker Miller and Governor Cuomo.[24] As a result, "The speaker wanted to stop high-tech pork" reports the budget official. Miller began the process that resulted in an Assembly bill introduced but not passed in 1988 to control such appropriations.[25] Also in 1988, Cuomo asked his lieutenant governor to review the selection of such projects.[26]

Completed in 1989, the lieutenant governor's report observed that state commitments for new science facilities had been made "in the absence of either a stated policy or formal programmatic context." Proposals had been evaluated by politicians and staffers who lacked experience in the sciences and technologies in question. "The majority of these extraordinary projects have been supported because of their professed economic development potential and opportunity," even though, as the report continues, such arguments were rarely made rigorous.[27]

Pork as Technology Policy

Lobbying organizations representing the country's large research universities vociferously condemned the trend toward earmarked funding, at least at the federal level. Their argument was that, by avoiding the scrutiny of research proposals by technical peers, earmarking undermined the quality of the country's technical research. Defenders of the practice, most notably Boston University's President Silber, countered that peer review restricted funding to elite institutions and prevented new centers of research from arising.[28] In agreement with Silber, RIT president Rose called traditional merit review an "old-boy network" that "stifles innovative research and contributes to stagnation in the nation's overall scientific effort."[29] He pointed out that, year by year, the twenty major research universities, the "haves," received 40–50 percent of federal academic research support, while the next five hundred "have-not" schools shared the remaining half.[30]

At least in the two instances cited here, the federal grants indeed allowed institutions with little funded research to build nationally recognized technical programs. Alabama's Center for Applied Optics, built at the Huntsville campus, and RIT's imaging sciences center may well be stories of success in the academic pork barrel. Both Huntsville and RIT were prepared to put extensive institutional effort into their new technical programs, possibly much more administrative effort than big-name campuses would have been willing to invest. With state, military, and industrial funding now supplementing federally earmarked funds, Huntsville has become a respected locale for optics research. RIT's new center built a faculty of eighteen highly regarded specialists in imaging, established the nation's first doctoral degree in imaging science, and attracted more than a dozen U.S. and Japanese corporate affiliates. Huntsville and RIT might have fared badly had they had to contend for peer-reviewed facility funding in a league that included, say, a Johns Hopkins or MIT. But in the world of earmarking, Huntsville and RIT might be exceptions. Furthermore, it is not at all evident that research quality is the main casualty in earmarked technology appropriations.

Since photonics is a reasonably obscure subject in most minds and not high on agendas of legislators, few institutions, perhaps just these two, acquired federally earmarked funds for it. Subjects such as agriculture, food science, cancer treatment, and "bioscience" were much more popular.[31] And earmarked projects varied in how they specified science facilities. At the extreme of specificity, the law set aside $3 million for "icing and related next generation weather radar and atmospheric research" at the University of North Dakota. At the other extreme, the law allocated "$650,000 for research and related purposes" at Iowa State, and $10 million for a Barry M. Goldwater Center for Science and Technology.[32]

Sometimes the wording in the appropriations law or in accompanying reports is so ambiguous that no one is sure what is being funded. "It is possible," writes Colleen Cordes of *The Chronicle of Higher Education*, "that both houses of Congress could earmark federal money for a university that only one lawmaker, or a few lawmakers from one state, could actually identify."[33] Such confusion seems to have occurred with a proposal for appropriations for an optoelectronics center. It was initiated by Senator Pete Domenici of New Mexico on behalf of the University of New Mexico. However, the legal language was so vague that the Defense Department was uncertain whether or not it was indeed obliged to forward the funds to the university.[34] After a dispute over the obligation between Domenici and Sam Nunn (chairman of the Senate Armed Services Committee), the Defense Advanced Research Projects Agency, which was slated to disburse the funds, finally concluded that it had the prerogative to distribute the funds through competitive applications.

The earmarked appropriations are frequently scattered through obscure legal documentation, so that no one can be quite sure how much earmarking takes place, much less discover the main topics of earmarking or enumerate the institutions to which it is directed. In an effort by the Congressional Research Service, one whose results are not definitive, researchers had to laboriously wade through numerous appropriations bills accompanying House and Senate conference reports, and public laws in order to identify earmarked college programs.[35]

The outcome of such scattered appropriations might be lowered research quality, or they might sometimes be opportunities for small colleges to build research quality. More importantly, the outcome is an implicit policy that has no coherent relationship to any public ends or to the critical needs of specific technological fields. In pork-barrel policy-making, decisions about the amounts spent on a research facility, the subjects to which it is devoted, and the facilities appropriate to conduct the research fall to individual politicians interacting with interest groups. The array of facilities so funded in any year reflects the push and pull of political bargaining, with no broader perspective on national or regional purposes. Even though the policies impinge variably on specific technological fields such as photonics, the policy-making incorporates no means by which the critical needs of the field can be understood.

The political forces favoring such policy-making have proved to be so enduring that by 1988 and 1989 the lobbyists representing the large research universities gave up on attempts for complete reform. The lobbyists now hoped at best for a compromise that would require peer-reviewed funding for most research facilities, while reserving funds for the less prestigious campuses. According to one lobbyist, "in raw political terms, you can't beat earmarking." The lobbyist added: "Once an earmarked project appears in public, it's practically impossible to turn around. The deals have already been made. These things become leverage items—they're currency in moving the larger legislation."[36]

Such deals require complex negotiations. Legislators or their staffs must find ways of agreeing on who gets how much for which project. In that convoluted task, the lobbyists have special capability. A Cassidy and Associates spokesman has been cited as saying that the company's clients have received 40 percent of earmarked university facilities funds since 1980.[37] Having so many academic institutions as clients, a lobbying firm would, as part of its service of setting up coalitions for clients, be in an exceptionally favorable position for facilitating the deals on who gets what.

National policy on physical facilities for photonics research, then, reflects decisions made by the interested parties themselves, the academic institutions searching for strategic resources by which to improve their prestige and standing. And such policy-making may also emerge as legislative compromises worked out, in part, through the services of a specialized lobbying firm.

BUSINESS PARTICIPATION AND MESUREMENT RESEARCH IN OPTOELECTRONICS

A brochure describing the work of the Center for Electronics and Electrical Engineering, a unit of the newly named National

Institute of Standards and Technology (NIST), heads a list of technical interests with "photonics," in which it includes lasers, optical fibers, and fiber-optic detectors and sources.[38] The agency's work in the field, though modest in cost and personnel, represents an element of the U.S. policy response to the commercial success of optical communications technology.

The National Bureau of Standards, as the agency was formerly known, has not, despite its name, engaged in setting or enforcing standards. By and large, such standard setting has been the province of hundreds of voluntary organizations. They are composed mainly of representatives from industry, with the bureau (now the institute) serving as one member among others. Rather than setting standards, the agency conducts research on metrology, the science of measurement. The agency has a role in photonics, therefore, not in setting standards but in the measurement research that makes standardization possible in optical communications.

The story of the agency's evolving interest in photonics illustrates how decisions were made to direct resources at a new technology and how the research agenda was set. Central to this decision making was diffuse business participation—amounting to less than active collaboration on joint committees—in the agency's practical decisions. This means of making technological choices was all the more significant because, in debates in the 1980s over government's proper role in building technological capability in industry, the bureau was often presented as a model of how government agencies should behave.

Since its establishment in 1901, the agency has undergone several shifts of its role in federal government. In the 1930s, before the enormous expansion of corporate research and government support for university research, the Bureau was often seen as the governmental locale for basic research for industry. Fulfilling a more limited version of this role in the late 1940s and 1950s, the bureau became the developer of demonstration projects and prototypes, until they could be taken over by industry. During that period, the agency developed one of the first operational computers, under contract from the Bureau of the Census and the Office of Naval Research.

The bureau's role in the late 1950s was to conduct research in physical measurement, this still seen as the agency's predominant purpose. In 1978 and 1979, during the Carter administration, the bureau's activities were further redirected, away from consumer products and the energy issues that occupied it in the mid-1970s, to matters of industrial innovation and productivity.[39]

As concern over foreign trade intensified in the 1980s, witnesses in congressional hearings repeatedly praised the agency as a model of proper relations between government and industry, and as a potential bureaucratic home for a range of programs meant to reassert U.S. technical strength in industry.[40] In 1988, when Congress gave the bureau its new awkward name and broadened the agency's mandate to include a range of new technology transfer programs, the stated purpose was to "augment [the agency's] unique ability to enhance the competitiveness of American industry."[41]

The congressional act explicitly retained the agency's traditional responsibilities in metrology. To this day, the agency devotes much of its effort to investigating the precise measures by which standard setting becomes possible. An industrial company, say in the fiber optics industry, needs the standards both to assure potential buyers that they are receiving products that conform to known criteria of performance and to allow the company to connect the fiber with light sources, detectors, modulators, and other paraphernalia used for optical communications.

The Origins of the Optoelectronics Program

Since the late 1970s, work on "lightwave technology," or "optoelectronics" has taken up a modest portion of the agency's resources. The work takes place within the Optical Electronic Metrology Group located in Boulder, Colorado, and occupies about forty researchers, including visiting researchers from industry.[42]

This present level of dedication to the fiber-optics industry evolved gradually from initial technical capability the agency acquired in the 1960s when developing safety standards for lasers and from work for the Department of Defense on procurement criteria for lasers. Having laser specialists who participated in professional conferences and standard-setting committees in optoelectronics, the agency was in a position to recognize evolving trends in the technology as early as 1976, when the decision was made to begin serious work on methods for measuring the characteristics of optical fibers. Since many of the bureau's scientists participated in standards committees organized by the Electronic Industries Association, and since most manufacturers of optical fiber and components were members, the association became the vehicle for industrial backing for the bureau's expanded work in fiber optics.

The bureau's role received more industrial support when, as staff members tell it, the two major optical fiber producers, Corning Glass and the then still-regulated AT&T, found they had irreconcilable techniques for measuring the characteristics of their fiber, as did various laboratories within AT&T itself. In the late 1970s and early 1980s, the bureau became the means by which the industry, especially the two largest domestic companies in it, could agree on common measurement techniques.

Through the early 1980s the bureau's work on the subject was funded only with a few hundred thousand dollars a year from the bureau's internal discretionary funds. But as multi-mode fibers (which carry numerous light signals simultaneously) were developed and came to use optical repeaters and optical switches, more advanced work on measurement devices and techniques would require a larger investment. The country was now being criss-crossed with long-distance optical telecommunications cable; the program would be serving a fast-growing high-technology industry.[43]

During this period, discussions began within the bureau on initiating a new budget item to cover optoelectronics. Budget documents were prepared for review by the bureau's parent agency, the Department of Commerce, and upon its approval, for congressional and executive-branch budget staffs. From 1983, the bureau's higher officials started approaching the executive Office of Management and Budget and congressional oversight committees for appropriations to directly support expanded work in optoelectronics.

In October 1985, the bureau's senior economist, Gregory Tassey, completed a planning report anticipating three stages in the evolution of optoelectronic technology. He forecast rapidly expanding worldwide research and commercial markets for the technology.[44] Also between 1983 and 1985, the bureau's technical staff and managers visited leading laboratories in Japan and the U.S. for updates on the state of fiber optic technology. They formulated a technical program for metrology in optical communications.[45]

The budget initiative proved to be successful in 1987, but barely so. Congressional support of optoelectronics metrology amounted to only $500 thousand in 1988. The fiscal year 1989 budget finally put the program on a firmer footing with a $2.4 million appropriation, including $300 thousand for continued laser work. The program also received about $500 thousand from the agency's discretionary funds, plus additional amounts received under contract with other federal agencies, for a total of around $4–5 million in 1989.[46]

What brought about this level of dedication to optoelectronics in 1989? As told by the agency's officials, the growth of the program reflected the staff scientists' own views of the industry as acquired through interactions within the profession. Studies showing evidence of the growing importance of the fiber optics industry bolstered this interest. But most of all, the growth of the program reflected the budget process and the strong interest shown by the Electronic Industries Association.

While still director of the National Engineering Laboratory, John Lyons (later director of NIST) recalled during an interview in late 1988: "I had to give testimony on photonics for years. We were developing the subject for five years before it got going. After you deal with a lot of groups for years, one day everyone agrees that this has to be done."

"We also do some in-depth studies," Lyons continued, "but formal planning documents are not really the process. We do them now and then when Congress asks. Most of the decisions are made through the budget process."[47] Gregory Tassey, the agency's economist, while stressing the role of economic studies, adds that "one has to sell the program initiatives and seek congressional approval. They're line items in our budget and are scrutinized at every step along the way."[48]

Setting the Research Agenda

How are decisions made at the agency on which lines of research to pursue? "There is no single way or single answer," answers one of the agency's officials, echoing several others.[49]

At a minimum, there must be interest by the staff researchers themselves. Robert Kamper, Director of the Boulder Laboratories, where NIST's optical electronics program is located, says that "the source of most new projects, certainly in our lab and probably everywhere else, is what we call the bench level." He adds that "for something to get started it requires one person who really wants to do it. Usually he gets his idea from talking to his colleagues in the field."[50]

Moreover, Kamper and all other agency officials who were interviewed stress that relationships with industry are central in decisions to pursue research. "We make our decisions on the basis of contacts in industry," states one official. "Their suggestions go a long way and weigh heavily on us."[51]

These relationships with industry take diverse forms. The agency's scientists meet with industrial counterparts at conventions, sit with them on standards committees, and visit industrial laboratories. The agency sends out publications and hosts conventions, such as annual ones on laser safety and optical-fiber measurement. Scientific "assessment panels" annually review the agency's work. What provides the closest contact is the Research Associate Program: visiting scientists from private industry, paid by their home firms, work for as long as two years in the agency's labs, mainly on problems of measurement research. The Optical Electronic Metrology Group hosts several such visiting researchers each year. The agency provides them with technical supervision, office and laboratory space, supplies, and the use of most research equipment, without charge.[52]

Business Participation and Public Policy

Since the National Institute of Standards and Technology has pursued research, albeit highly specialized research, meant to answer to industrial needs, the agency has long enjoyed diverse and diffuse relationships with industrial firms. In a time of foreign industrial ascendance, the agency's officials have lost no opportunity, especially in congressional hearings, to describe the tradition of close contact with private firms.[53]

In one of several sets of hearings leading up to the agency's 1988 name change and expanded mission, comments by Lewis Branscomb, the agency's former director and former IBM chief scientist, now a professor at Harvard, were especially pointed. Branscomb came to address the House Subcommittee on Science, Research and Technology in 1987, as he put it, "to speak on the place of the National Bureau of Standards in the nation's competitiveness strategy."[54] He told the congressmen:

> The Bureau also, as I mentioned before, has an excellent way of working with industry. Sufficiently excellent that you will never hear, at least I have never heard, of the Bureau of Standards' work with industry referred to under the phrase, "industrial policy." I think that's another thing we want to keep in mind, if industry is to benefit from the Bureau's work, it's very important that the relationship be one which industry finds supportive and helpful and not threatening.[55]

In his prepared statement, Branscomb reiterated: "Critics of 'industrial policy' can cite the NBS mode of operation as an exemplar of the right federal posture."[56] Indeed, federal legislation of the 1980s seemed intent on having other federal research agencies benefit from the example. With passage of the Technology Transfer Act of 1986, each federal laboratory employing two hundred or more scientists or engineers was required to set up an Office of Research and Technology Applications with the function, among others, of arranging cooperative research between the laboratory and industrial firms.[57] Further legislation specifically required the Department of Energy's laboratories, many operated by contract organizations, to engage in technology transfer and cooperative research activities with business.

By 1991, Lawrence Livermore National Laboratory (LLNL), a major weapons lab, established an office for the application of laser and electro-optic technologies. According to the director of this office, federal policy was encouraging the laboratory "to work consciously and deliberately with the electro-optics industry to establish working partnerships to commercialize LLNL technologies."[58]

Diffuse Decision Making and Missed Opportunities

The enthusiasm about business participation overlooks a study suggesting that the standards agency's decision making by such means has its pitfalls. Prepared by the Congressional Research Service in 1981, the study observes that, "on the one hand," the hundreds of standards groups and technical panel meetings in which the bureau's staff participates "are the traditional ways in which research scientists and engineers present and test their ideas and sort out their priorities." "On the other hand," the author adds, "several present and former NBS officials spoke of 'missed opportunities' for initiative and leadership in areas that developed in the 1970s."[59]

An example of a missed opportunity in the 1980s might be electronic imaging, which combines devices that capture images, turn them into electronic signals, process the signals, store them, display the image, and allow users to manipulate the image and to print it. Since these functions are performed by discrete hardware devices and software programs, each probably made by a separate company, they require calibration, standards, and protocols to be integrated into workable imaging systems.

Indications in the 1980s were that the production of imaging devices was becoming an immense industry, but that judgment would have been hard to make. No disinterested agency systematically tracked the development of such new clusters of industry and the privately produced reports that were available sometimes acted as enthusiasts for a technology. By the mid-1980s, analysts at financial houses in New York were contending that electronic imaging was a promising, fast-growing field.[60] One consulting group's study showed the imaging industry valued at $5 billion by 1988. The report concluded, however, that "until device specifications and interface standards are developed and distributed, explosive growth for EI [electronic imaging] is extremely unlikely."[61]

But the conditions that permitted the agency to respond early to fiber optics were absent in the case of imaging. At least through the mid-1980s, most imaging companies were small[62] and had no well-developed industrial association. As compared to its experience in fiber optics, the National Institute of Standards and Technology had fewer laboratory facilities and staff members devoted to work on imaging devices. And there were competing technological developments, such as those in advanced ceramics and superconductivity, pressing for the agency's attention.

Broadly connected to the private sector through a variety of industrial contacts, the agency's management and staff members did recognize the importance of the field. Indeed, outside observers frequently complimented the agency for being well aware of current developments. But could such recognition produce the industrial backing and political support needed to obtain funding for a concerted laboratory effort in metrology?

By 1991, several small research projects related to imaging were scattered among the agency's labs, including the laboratories of electronics and electrical engineering, computing systems, and applied mathematics. Additonal projects related to imaging were being started up that year with the backing of the defense agencies, including the Defense Advanced Research Projects Agency.[63]

Also by that year, NIST's director testified to Congress that U.S. industry was pursuing imaging technology vigorously. He said, "The American Electronics Association, for example, has scheduled special meetings with me and other NIST staff to assure that we are aware of the range of private activities under way."[64] But it would still take several years to consolidate scattered projects and build up the staff and laboratory facilities needed for a major program of measurement research. The agency would have to work with industrial supporters, officials of the Commerce Department, and budget staffers to obtain a specific appropriation.

In the case of fiber optics, after several years of politicking by the agency's highest officials, such diffuse interactions did eventually yield a significant research program. But this outcome depended on a fortuitous combination of events. Staff capability had been built from earlier work on laser calibration and electronics. Two large corporations, AT&T and Corning, were planning to supply or install thousands of miles of optical cable. And the Electronic Industries Association was ready to serve as an organizational vehicle to represent the firms. Without such ingredients fortuitously combined, a set of industrially significant technologies might fail to receive the agency's concerted attention just in their emerging stages, when a domestic industry is most vulnerable to developments in other countries whose technology policies are more coherently organized.

LIGHTWAVE TECHNOLOGY AND MERIT REVIEW AT THE NSF

By the latter half of the 1980s, several branches of photonics received substantial support at the National Science Foundation. During the 1988 fiscal year, the foundation devoted an annual $13 million to research in photonics. When an NSF official inventoried the Foundation's support for photonics research that year, he found that most of it was in the Engineering Directorate. Within the directorate, much of the work on photonics took place through a new program that designated collaborative centers for engineering research at universities. (The program is discussed in chapter 6.) The directorate's more traditional funding programs distributed research funds to individual investigators. Two programs did so for photonics. The Quantum Electronics, Waves, and Beams Program devoted about $2.2 million, half its budget, to research in this field in 1988. The directorate's Division of Emerging Engineering Technologies included a Lightwave Technology Program, which was entirely directed at this subject. The program was responsible for awarding $3.1 million in 1988.[65]

The Origins of the Lightwave Technology Program

The establishment of the Lightwave Technology Program and of most of the rest of the engineering directorate's work in photonics research date from 1985, when the directorate underwent extensive reorganization. Amid the reorganizing, a new Division of

Emerging Engineering Technologies was created. How did lightwave technology come to be chosen as one of the emerging technologies to be included in the new division?

As the story is told by present and former NSF officials, the program in part continued earlier work in a previous organizational format. Research in optical communications started at the foundation in 1971 and evolved into an "organized area of interest" pursued modestly through several NSF funding programs. Devotees of the subject at NSF organized periodic meetings at universities to bring together investigators working in the field. In February 1984, an NSF-sponsored workshop entitled "The Future of Lightwave Technology" was held at the University of Southern California and drew 160 participants.[66] The origins of the Lightwave Technology Program at NSF have been traced to this meeting.[67]

In the same year, 1984, grander Reagan administration politics were at work to reshape the Engineering Directorate. (The story of that reorganization is pursued in a description of NSF's engineering research centers in chapter 6.) Also in 1984, the directorate became the responsiblity of a newly appointed assistant director of NSF. In keeping with NSF's practice of bringing academic specialists to administer its programs, the new assistant director, Nam P. Suh, came to the agency temporarily from his post as a professor of mechanical engineering at MIT. Decisions made during his tenure at NSF led to the creation of the new Emerging Engineering Technologies Division, of which the Lightwave Technology Program was to become a part.

Suh entered his new office in Washington with convictions about the directions engineering should take in this country. In recalling his reasoning, Suh explains that "NSF is not an entitlement program for universities," but is charged by Congress with larger roles of contributing to scientific and engineering progress, national defense, and prosperity. In Suh's view, NSF had traditionally allowed pressure from proposals to drive its programs in engineering, and consequently the agency could not provide leadership in developing new technological fields. NSF in effect supported research in established disciplines at the expense of new subjects of engineering that crossed disciplinary lines. In creating the new Division of Emerging Engineering Technologies, says Suh, "my idea was to develop a new division at NSF in order to help universities move into new areas."[68]

Suh takes as an example optoelectronics. University research in the subject is typically scattered among electrical engineering, materials engineering, physics, and other departments. "So," he continues, "the question is how to develop future optoelectronics specialists who can then become champions of new fields, allowing academic institutions to develop new programs."[69] He and his colleagues chose biotechnology, neuroengineering, and other fields, along with lightwave technology for the new division because these were engineering subjects crossing traditional disciplinary boundaries.

But Suh had to mobilize support for such changes in funding programs at NSF. Between the time he was contacted by the White House in April 1984, until FBI checks were complete and he was confirmed by the Senate in October, he worked to garner support for the changes. He met with staffers at the Senate, House, and Office of Management and Budget, held meetings with the National Academy of Engineering, and gave speeches at professional engineering meetings. In Suh's view, he succeeded in gaining support for the new division of emerging engineering technologies because he had "done a lot of legwork beforehand."

In choosing fields to be included as emerging technologies, "there was no set formula," explains Suh.[70] During the reorganization, he did establish a "strategic planning process," pursued by internal staff groupings. Referred to as "task forces," they brought together staff members to discuss the directorate's goals in a time of transition and the form that reorganization would take. Task forces also offered a way of building staff support for the proposed changes.[71] Suh remarks that he favored more extensive planning, but such strategic planning that he did institute raised a great deal of opposition. Its opponents, according to Suh, "said Congress should do planning, not agencies."[72]

Paul Herer, who worked closely with Suh in the reorganization, also recalls that "We did not document a substantial justification for the programs... I've found that priority setting and the way an initiative gets proposed doesn't always occur the same way and with the same procedures." Some budget proposals entail extensive discussion and justification. But "sometimes it seems that an assistant director just says he wants it and puts it in the budget documents."[73]

While Assistant Director Suh did reorient the directorate partly through his own decisions on what was better, he went through the politicking needed to gain budgetary approval and made sure that the parties most interested in the engineering grant programs, mainly the academic engineering educators, were substantively in agreement with his proposals. To the extent that the establishment of the new Division of Emerging Engineering Technologies, and the Lightwave Technology Program within it, represented his own vision, it became susceptible to rearrangement by others. As of 1989, his successor had refolded emerging engineering technology programs into fields associated with more traditional academic disciplines. The lightwave technology was placed under the wing of the Quantum Electronics, Waves, and Beams Program.

The Review Process

In 1985 the Lightwave Technology Program was set up as one of the engineering directorate's programs in emerging engineering technologies. In keeping with the preponderance of NSF activities, the program distributed its funds to individual researchers and research teams at universities around the country. The program's budget of $3 million, which remained fairly constant through 1989, was adequate to fund twenty to thirty research projects a year, and most of it went to the continuation of multiyear projects that had originally been awarded funds in a previous year.[74]

How are decisions made on awards for the ten or so new grants awarded each year? The decisions are made through peer review, or rather, through "merit review"—a term adopted at NSF in 1986. Since the name change occurred in the wake of a debate about the efficacy of peer review as a means of making technology policy decisions for public ends, the story of the name change can illustrate the dilemmas of a colleagual decision procedure in technology policy.

For the most part, merit review operates as peer review previously did. Projects are selected by a group of technical specialists who review written proposals submitted by prospective principal investigators. In larger NSF programs, such as the one that selects engineering research centers, the review is conducted by a panel of experts brought together in Washington to look over a pile of proposals. For smaller programs such as lightwave technology, the program director mails out proposals, along with standard instructions and evaluation forms, to a number of reviewers who rate the proposals and whose identities are not revealed to the applicants.[75]

At the Lightwave Technology Program, and at the Quantum Electronics, Waves, and Beams Program (which also devotes most of its funds to photonics), the program manager generally mails his proposals to six reviewers and requires at least three properly prepared reviews to make a decision. Unless the reviewers are unanimous in indicating an exceptionally fine or very poor proposal, the program manager retains considerable discretion in choosing among the proposals. In making their choices, the program manager and reviewers are, by NSF rules, bound to consider proposals according to criteria specified on the standard evaluation form. The changes in these criteria are what mark the difference between peer review and merit review.

From Peer Review to Merit Review

The changes in the criteria, though not in the name, date to the mid-1970s, when NSF's program of Research Applied to National Needs (a division that was soon to be dissolved) brought out new criteria that incorporated thoughts of social benefit in project selection. The new criteria took their place on NSF evaluation forms without incident through 1985, when the foundation was forced to look again at its project selection procedures.

The main incentive for the review of the procedures was the controversy that arose over congressional earmarking of funds for academic research facilities. Defenders of earmarking contended that NSF's selection procedures were biased in favor of a few regions, well-established institutions, and well-known professors. In response, the National Science Board, a board of advisors to the foundation, recommended to the NSF's director that he appoint a committee to evaluate project selection procedures. The resulting NSF Advisory Committee on Merit Review issued a report the next year.

However, the committee went out of its way not just to answer critics of distributive equity but also to look at the usefulness of research selection procedures for larger national ends, primarily competitiveness and defense. The committee stated in its report that the foundation's selection procedures were in fact not narrowly based on peer review. They were already taking broad social goals into consideration. In the committee's view, the "relevance criteria" of the 1970s proved to be flexible enough to serve the new national R & D priorities of the 1980s, including regional equity and "the development of key national needs for defense and for economic competitiveness."[76]

Hence the committee made it its primary recommendation that no change in selection procedures was needed. Rather, the term *merit review* should be adopted to recognize that selection criteria include "not only technical excellence but also additional factors."[77] In his preface to the committee's report, Erich Bloch, director of the foundation, stated his acceptance of the name change and directed the agency's officials to adhere to it.

Grants Awards "in the Solution of Societal Problems"

Reflecting these broader research aims, the proposal evaluation form used by the Lightwave Technology Program instructs reviewers to consider a proposal according to four evaluation criteria. The first and second instruct reviewers to assess the technical soundness of the proposal, the capability of the investigators, the "intrinsic merit of the research," and the likelihood that it will lead to new discoveries. The fourth criterion asks reviewers to look at the "effect of the research on the infrastructure of science and engineering."

The third criterion, entitled "Utility or Relevance of the Research," is the broadest. According to the form, the criterion is to be stressed in the evaluation of applied research projects, which would include the projects of the Engineering Directorate. The third criterion reads as follows:

> Likelihood that the research can contribute to the achievement of a goal that is extrinsic or in addition to that of the research field itself, and thereby serve as the basis for new or improved technology or assist in the solution of societal problems.[78]

In just one year (1985), almost sixty thousand NSF panelists and mail-in evaluators reviewed 24,403 proposals.[79] A significant number of the projects were applied projects evaluated by the Engineering Directorate. In such applied projects, the third criterion would have to be taken into consideration in project selection. NSF officials who were interviewed could not, however, recall instances where reviewers were unduly daunted by the prospect of deciding whether a research project would assist in the solution of societal problems.

The Advisory Committee on Merit Review did recognize that reviewers selected for their technical knowledge "are not necessarily best equipped to make judgments about the potential utility of research, or its impact upon the infrastructure of the science and engineering system."[80] Outside observers have also been skeptical of the use of peer review for priority-setting, pointing out that reviewing research proposals for "truth" is not the same as reviewing them for "utility." Peer scientists have doubtful capacity for the latter.[81]

The advisory commmittee held, nevertheless, that most NSF programs that have utility as one of their goals include in the process reviewers who have experience with such goals. The committee gave only one example. It stated that "engineering programs, for example, have long used industrial reviewers as a matter of course."[82] Though the committee did not go into more detail, it seemed to hold the position that reviewers' industrial experience would be the source of knowledge from which to judge the usefulness of research projects for greater public purposes.

Grant programs at the engineering directorate, including the Lightwave Technology Program, did indeed use industrial reviewers, though the number of mail-in reviewers from industry was not recorded. The foundation refrained from surveying its reviewers to avoid risks of compromising their anonymity. The manager of the lightwave technology program estimated that about one third of his program's reviewers were from industry.[83]

Also, the program announcement for lightwave technology encouraged applicants to identify research projects in collaboration with other institutions, including industrial organizations.[84] In the words of Lawrence Goldberg, manager of the Quantum Electronics, Waves, and Beams Program, "we look to increase connections between academy and industry." "Generally," he added, "I encourage applicants to build into proposals relationships to industry as a way of strengthening proposals to pass through the review process."[85]

Other than by involving industrial scientists and engineers in the selection procedure, the NSF programs instituted no means by which reviewers were to grapple with the broader social and economic ends that they were now charged with evaluating. Traditional peer review had selected projects through a process that required reviewers only to make contextual judgments about their own technical field. But merit review asked them to select projects in lightwave technology by judging broader effects on engineering "infrastructure," and to evaluate other social ends, though no intellectual means for making such judgments had ever been articulated. In effect, the merit review process drove public decisions down to anonymous committees that had no discernible means of making the judgments for which they were formally responsible.

CONCLUSIONS: DISAGGREGATIVE POLICY AND PHOTONICS

A range of policies in the 1980s sought to reassert national and regional technological strength through policies of making government resources available for sector-specific research programs, including those directed at photonics. In the midst of the widespread opposition to explicit industrial policy, sector-specific programs operated in disguised forms. Technological programs for industrial purposes operated as we shall see later, under the auspices of military agencies or through collaborative committees of industrial and academic representatives. But in the most traditional form of governmental response, disaggregation, resources were distributed unit by unit with no consideration of a broader strategy or plan. In disaggregative policy, governmental assets were made available through pork barrel, business participation, and merit review.

As part of the rapid rise of federal and state pork-barrel appropriations for high-technology economic development in the 1980s, photonics research facilities were funded at a few campuses in New York, Alabama, and perhaps elsewhere. The method of selection was back-room budgetary negotiations in which specialized lobbyists had a leading role. Once the earmarked science facilities were built, the only federal requirement was that the facility indeed fulfill any function implied by a four- or five-word mention in a legislated budget document. The choice of activities within the facility was the prerogative of the host institution.

Over the same decade, optoelectronics programs were significantly strengthened at both the National Institute of Standards and Technology and the National Science Foundation. A core of interested employees at the institute and an administrative champion in the foundation provided internal support within the agencies. Lobbying by the agencies proved essential to pass the program through the budget process. And the interested parties had to give their backing, the fiber-optics industry in one case and the academic engineering community in the other.

At the National Institute of Standards and Technology, the optoelectronics research agenda and facilities were made available for the use of business participants. At NSF, decisions on the distribution of research funds for lightwave engineering were placed in the hands of anonymous university and industry reviewers. NSF requirements made the reviewers formally responsible for considering public benefit in their choice of projects. The anonymous reviewers made technological decisions for public ends without engaging in any deliberative process or giving an intelligible accounting for the reasons by which they made their judgments.

In combination, these kinds of disaggregative policy made available significant federal assets for photonic research in the 1980s. Decisions on where to direct this national investment took place through political and legislative deals guided by specialized lobbyists, administrative and staff decisions to build internally favored programs, and the give and take of the budget process. Moreover, the interested parties themselves helped determine the research and funding choices through lobbying on behalf of the programs, participation in government laboratories, and an anonymous technical-review process.

The sum of these decisions constituted a privatized industrial policy operating in disaggregative form. In this form of privatization, policy-making directed toward photonics was pushed out to the interested parties and negotiated against other interests through back-room give and take, as if such accretions of discrete self-interested decisions could respond to the moral and industrial implications of a technological revolution.

SIX

Collaboration

WHAT'S NEW ABOUT COLLABORATION?

Policies of distributing government assets to private interests through tax incentives, business participation, and peer review were the older fashion in American technology policy. Policies of encouraging research collaboration among firms or between industry and universities were, however, a model of the 1980s. Research centers initiated through state or federal policies and operated through collaborative committees would now be the primary institutions of U.S. civilian technology policy. Since photonics achieved its industrial importance in the same decade, various programs of joint research in advanced optics serve to illustrate the new collaborative ethic.

The collaborative response to U.S. industrial problems occurred in two variations, one being the industry-university joint research center. Operating under the auspices of a university, the research centers allowed fee-paying companies to take part in research decisions and gain privileged access to research results.

Though some research centers were initiated with universities' internal funds, most of the larger ones arose through governmental encouragement and funding. Policymakers saw in the research centers a way to avoid the dispersion of research funds to individual investigators. Policymakers hoped, rather, to create productive interrelationships among researchers in one institution and encourage these researchers to allocate their funds according to some sense of economic relevance. The knowledge by which university research centers were to undertake this allocation was to be acquired through collaborative relationships with industry. But collaborative arrangements left the research center open to the contrary pressures of, on the one hand, industrial affiliates who wanted to use the research center for cut-rate contract research and, on the other hand, other corporate affiliates and faculty members who wanted to maintain professorial control over the research agenda. Caught between these tensions, the research centers never developed the means to allocate research projects according to a more coherent notion of industrial and public ends.

The collaborative response of the 1980s also occurred in another variety, the industrial research consortium. Such a consortium can be defined as a free-standing organization established by private firms, even direct competitors, to conduct research (or fund research) for the benefit of member firms. In the field of photonics, though industrial consortia were frequently discussed, only one came into being and it survived only for two years. In photonic research, university-industry collaboration proved to be much more popular. Chapter 6 is devoted primarily to the industry-university brand of collaboration. But since the industrial research consortia were a prominent form of privatized technology policy in other technical fields, they deserve further comment.

Industrial Consortia

Industrial consortia appeared soon after the turn of the century, when trade associations set up research institutes in electrical-lamp making, canning, horology, and brewing. But strengthened antitrust laws, including awards of triple damages for certain violations, inhibited many kinds of technological cooperation.[1] In the early part of the 1980s, by far the most significant examples of collaborative industrial research were to be found in regulated industries whose firms did not directly compete with each other. Of an estimated $1.6 billion expended on collaborative R & D in 1984, about $1.4 billion of it was spent by groupings of regulated firms: Bell Communications Research, the Electric Power Research Institute, and the Gas Research Institute.[2]

Then, in 1984 unanimous action in both houses of Congress eliminated most of the antitrust restrictions on such research relationships. The National Cooperative Research Act protected R & D consortia from antitrust suits unless restraint of competition could be shown in practice. This act also eliminated the threat of treble damages.[3] Within three years, sixty-seven consortia registered with the Department of Justice under the act in research fields ranging from software engineering to magnesium development.[4]

The commentary on legislative encouragement of the consortia was clear: the federal government supported them with the hope that they would be good for competitiveness.[5] The new consortia arose, then, through public policy that meant to have industrial cooperation play a role in the U.S. response to foreign technological rivalry. The creation of consortia reflected in clear form a policy in which technological decisions were relegated not to the market but to private organizations that would coordinate private actions for national economic purposes.

Of the new consortia, several registered their intention to pursue research in optics or optoelectronics. In seeming disregard for the spirit of the law, however, most of these arrangements were simply research ventures between two firms, often with one of the pair being a foreign firm.[6] Only one consortium in photonics, the Battelle Optoelectronics Group, truly conformed to the spirit of the legislation in having several domestic members and a broad research agenda.

Registered in November 1985, the six corporate members operated under the wing of Battelle Columbus Laboratories, a contract research organization in Ohio. The consortium aimed to investigate manufacturing and packaging techniques for the optoelectronic components used in fiber-optic communications. Each corporate member (including a seventh firm that joined in 1986) paid an annual fee of $200 thousand. Its former director, Robert Holman, stresses that qualities of leadership and charisma, a carefully designed contract that spelled out rights and obligations of members, and the expectation that there would be direct payoffs to member companies were all critical ingredients in the establishment of the organization. But the consortium dissolved in two years. Holman attributes this partly to events in the administration of Battelle.[7] So, while rather large consortia in microelectronics, semiconductor materials, and computer technology managed to survive through the end of the 1980s, no substantial and enduring consortium emerged in photonics.

Consortia in all fields frequently proved to be troublesome. One diagnosis was that corporate participants sought to extract the most from the consortium while contributing the least. They faced problems of the logic of cooperative action, where each member acted strategically toward the others. Asked to comment on the advantages and disadvantages of collaborative research arrangements, the research director at Xerox Corporation expressed the fundamental problem as follows:

Arrangements are difficult because each participant or firm tries to get maximum benefit while minimizing the amount of commitment it makes. It's elusive how to make them successful. You're faced with how to get the most out of it with putting the least in.[8]

An analysis of one of the most prominent consortia, the Microelectronics and Computer Technology Corporation, similarly found that one of the fundamental problems in its operation was governance. The authors write that the consortium, seen as an institution, faced severe problems of internal governance, in contrast to foreign research consortia that were "nurtured and disciplined by public policy."[9]

One may speculate from such observations that the weaknesses of the consortia derive from the very process of privatization: the loss of authority that occurs when government relegates its policy-making capacity to private interests. One should, however, arrive at such a conclusion through a study of several collaborative organizations, a study that would have to expand beyond photonic technology.

The other form of research collaboration that arose in the 1980s, university-industry research centers, proved to be a more popular and enduring vehicle for photonics research in the United States.

The Proliferation of Industry-University Arrangements

Between 1980 and 1991, more than twenty industry-university collaborative research centers in photonics were established in the United States with state or federal funding (table 6-1). Each one was established with a unique financing package from sources that often included the host institution and private firms as well as government. Yet the governmental role was primary, because state or federal funding frequently served as the main inspiration for others to join. Added incentive came from federal tax deductions for contributions to basic research and from federal legislation that gave universities the right to own (and license out to firms) patents developed with federal funds. University-industry collaborative research centers were, to a substantial degree, the product of public policy.

In federal civilian government, the greatest originator of collaborative arrangements was the NSF, which established four collaborative centers in photonics by 1988. Research centers

6-1
University Research Centers in the United States in Photonics

University and Center	Year Founded
Alabama at Huntsville, University of (Huntsville, Ala.)	
Center for Applied Optics	1985
Arizona, University of (Tuscon, Ariz.)	
Optical Circuitry Cooperative	1984
Optical Data Storage Center	1986
California at Santa Barbara, University of (Santa Barbara, Calif.)	
Center for Quantized Electronic Structures	1989
Carnegie Mellon University (Pittsburgh, Pa.)	
Center for Excellence in Optical Data Processing	1983
Central Florida, University of (Orlando, Fla.)	
Center for Research in Electro-Optics and Lasers	1980s*
Colorado, University of (Boulder, Colo.)	
Optoelectronic Computing Systems Center	1985
Columbia University (New York, N.Y.)	
Center for Telecommunications Research	1985
Dayton, University of (Dayton, Ohio)	
Center for Electro-Optics	1983
Illinois, University of (Urbana, Ill.)	
Engineering Research Center for Compound Semiconductor Microelectronics	1986
Michigan, University of (Ann Arbor, Mich.)	
Center for Ultrafast Optical Science	1991
New York Institute of Technology (Glen Cove, N.Y.)	
Center for Optics, Lasers and Holography	1980
Polytechnic University (Brooklyn, N.Y.)	
Institute for Imaging Science	1981
Princeton University (Princeton, N.J.)	
Photonic and Opto-Electronic Materials Center	1989
Rochester Institute of Technology (Rochester, N.Y.)	
Center for Imaging Science	1985
Rochester, University of (Rochester, N.Y.)	
Center for Advanced Optical Technology	1983
Center for Optics Manufacturing	1989
Center for Photoinduced Charge Transfer	1988
Rutgers University (Piscataway, N.J.)	
Fiber Optics Materials Research Program	1985

6-1 *(continued)*

University and Center	Year Founded
Southern California, University of (Los Angeles, Calif.) Center for Photonic Technology	1980s*
Texas at Dallas, University of (Richardson, Tex.) Center for Applied Optics	1985
Virginia Polytechnic Institute (Blacksburg, Va.) Fiber & Electro-Optics Research Center	1986

* Year of establishment was not available

Source: International Society for Optical Engineering, *Optics Education 1989* (Bellingham, Wash.: SPIE, 1989); National Science Foundation, *NSF Science and Technology Centers*, 1991; *Research Centers Directory* (Detroit, Mich.: Gale Research Co., 1989); and partial telephone verification, 1989.

established through federal earmarking, such as RIT's Center for Imaging Science, also took on a collaborative university-industry structure. Defense agencies dabbled in collaboration, setting up collaborative centers for optical manufacturing at the University of Rochester and optical computing at the University of Alabama at Huntsville.

State governments were prolific in setting up research centers. Among the states, a new breed of science and technology agencies (or an older species, earmarked appropriations) established photonics research centers in Arizona, Colorado, Florida, New Jersey, New York, and Virginia. And universities also used internal institutional funds to set up dozens of small laboratories in optics, imaging, and optoelectronics, generally hoping to use them as leverage for further corporate and governmental support.[10]

Reasons for Collaboration

The new industry-university arrangements came into being over the same years that there occurred a remarkable number of conventions, conferences, and collections of papers on the values of intersectoral research cooperation.[11] Corporate reasons for the enthusiasm included an interest in radical new technologies, such as biotechnology, in which technical capability resided mainly on the campus; a sense that technical change was proceeding at such a rate that even the largest corporations could not keep up just with in-house resources;[12] and a response to tax incentives and implicit subsidies that went along with governmental support for the new arrangements. Furthermore, when research occurred under the auspices of university faculty, and did not try only to accommodate to the particular needs of fee-paying firms, the participating companies did not encounter the problems of cooperation as severely as they did in industrial research consortia.

Universities, for their part, sought the new research funding at a time when federal research support was leveling off. And all sectors—government, industry, and university—hoped to emulate what was seen as Japanese intersectoral cooperation.

Most of all, the enthusiasm for the new arrangements occurred in an atmosphere of concern about competitiveness. Browsing through the proceedings and testimonials, one finds that no values are more commonly expressed than those of the role of technology and cooperation in industrial competition. An example can be taken from the words of Dale R. Corson, former president of Cornell University, who served as first chairman of the Government-University-Industry Research Roundtable, one of a number of groups of academic, governmental, and corporate leaders that set out to foster industry-university arrangements. He remarked as follows to the roundtable's members:

> Competitiveness is currently all-important in our society. . . .
> How we approach competitiveness colors everything in university-industry alliances. The productivity of the research enterprise is one of the central elements in the portfolio of requirements for maintaining the international competitiveness of the U.S.[13]

Industry-university research arrangements were themselves not new, having a history that went back to the early part of the century. Industry-university cooperation can be traced back as far as 1907 at the University of Kansas and 1913 at the University of Pittsburgh, where manufacturers set up programs in which they supported personnel and defined research work. According to historian David Noble, the Massachusetts Institute of Technology became the first institution to thoroughly formalize industrial involvement in its research.[14]

The collaborative industry-university research centers of the 1980s are, however, not a simple continuation of earlier arrangements. An indication of change is that during the 1980s they suddenly became far more numerous. According to the proceedings of one conference held by the Government-University-

Industry Research Roundtable, "virtually all knowledgeable people who have commented on the matter have remarked on a virtual explosion over the past several years in the number and variety of university-industry alliances."[15]

More important than the increasing number of such arrangements were their new intent, new organizational forms, and implicit function of making allocative decisions for public ends. By the 1980s, the new research centers were not simply meant as short-term arrangements of convenience between individual companies and academic engineering departments. The industry-university research centers, as well as the new industrial research consortia, were the outcomes of public policy that intended to make them the cornerstone of a national (or regional) technological response to industrial rivals.[16]

The research centers were also seen as a new organizational form through which research would become more productive. By focusing technological skill in one place, universities and industries could attain greater economies of scale in research. Clusters of complementary skills would emerge that might not when research funding was distributed to individual investigators around the country. Researchers could more easily explore new technical fields not well covered within traditional engineering disciplines and could more unreservedly engage in projects crossing the boundaries between traditional engineering fields. National and regional centers of research strength would avoid the wasteful duplication that occurs when a technical field is pursued at dispersed sites.

Students would receive advanced technical training through an exposure to the broad technological ramifications of a new field of inquiry. In the field of photonics, research centers could afford specialized equipment, such as expensive photolithography devices, electron microscopes, optical fabrication facilities, and laser systems, some of which would otherwise be beyond the means of most college labs. Concentrations of advanced education, technical skill and costly equipment would yield, in the terminology of the early 1980s, "centers of excellence," which would achieve a critical mass allowing for more productive technical innovation.[17]

Inherent in such thinking, but inexplicit, was the idea that the technical investigators in such centers would become better attuned to the needs of industry and set their research agendas more wisely. Technical progress would be less dependent on the signals given by awards of grants, academic prestige, and corporate research rivalries. It would occur, instead, more through mutual consultation, reflection, strategy-making, and cooperative priority-setting on the development of a technical field.

Since the research centers came to be established through government policy and were intended to resolve national industrial problems thought to have a technological origin, they had bound up in them a further and even less explicit function. They would, in effect, be making research decisions for public ends. Policy and planning choices would be inherent both in the choice of the subjects in which the centers would specialize and in the setting of each center's research agenda.

How indeed did each research center come to set this agenda? The research centers' directions were supposed to become attuned to industrial needs, and research priorities were to be set—in short, industrial policy making would occur—through collaboration with industry.

The sections that follow examine the origins and outcomes of the collaborative industry-university research movement in photonics. The next two sections trace how the policy evolved and photonics research centers were selected at the National Science Foundation and in state governments, while keeping the government agencies out of agenda-setting roles. The sections that follow look at how research centers devoted to photonics were typically organized and at the subtle balances of power through which they set research agendas. The chapter concludes by recapitulating the characteristics of collaborative research as a form of privatized technology policy.

THE CREATION OF THE NSF'S ENGINEERING RESEARCH CENTERS

The NSF's Collaborative Research Programs

The reorganization of the Engineering Directorate, which created the Lightwave Technology Program in 1985, also installed the Engineering Research Centers. The creation of these and other similar NSF programs marked a sharper departure from traditional NSF practices than did the setting up of the programs for emerging engineering technology. Established at selected universities, the centers were intended to bring together engineering researchers across disciplines to investigate subjects of industrial relevance. Though the programs were set up with the purpose of contributing to industrial competitiveness, their subjects of research and means of setting priorities were left to industry-university groupings with no federal role, except that of assuring research quality.

The Engineering Research Centers program was one of a number of NSF attempts to create concentrations of research strength in fields of science and engineering. The engineering centers were administratively separate from other programs with similar names, including the Science and Technology Centers and the University-Industry Research Centers (which were specific NSF programs that should not be confused with industry-university collaborative research centers in general).

Launched as a pilot program in 1972 and implemented broadly during the Carter administration, the University-Industry Research Centers Program came by 1987 to include thirty-nine centers in subject areas ranging from hazardous waste to lymphocyte technology.[18] Among them was the Optical Circuitry Cooperative at the University of Arizona. Established in that university's Optical Sciences Center in 1984, it began its first year with $150 thousand from NSF, $220 thousand in the first of continuing annual appropriations from the State of Arizona, and $400 thousand from industrial membership fees.[19] As in the research centers that would follow, industrial collaboration was an essential element in the center's operations. The industrial members were to be represented on an industrial advisory board.

The new program of Engineering Research Centers differed from the University-Industry program in a few respects, the most important being scale. The new program funded engineering research centers at up to $2 million a year, as much as fifteen times that provided by the older program.

The Engineering Research Centers themselves inspired a still larger program announced by Ronald Reagan in his 1987 State of the Union Message in which he directed the NSF to stress "fundamental science that has the potential to contribute to our nation's economic competitiveness." Political commitments made after the speech supported the doubling of the agency's budget over the next five years. Among the centers set up in the early years was one devoted to the optical and electronic properties of microscopic atomic structures (at the University of California, Santa Barabara) and another to the study of extremely fast optical pulses (at the University of Michigan). Since these research sites were expected to direct their work at basic scientific questions, industrial participation was encouraged but not mandatory.[20]

Through 1990, the engineering research centers remained the most important example of civilian federal funding for collaborative university research directed at industrial needs. Of eighteen engineering research centers designated by 1988, three would come to devote substantial portions of their work to photonics. The Center for Optoelectronic Computing Systems jointly run in Boulder by the University of Colorado and Colorado State University was fully committed to optoelectronics. The Center for Compound Microelectronics at the University of Illinois at Urbana-Champaign devoted itself to the practical realization of optoelectronic integrated circuitry. And the Center for Telecommunications Research at Columbia University conducted an important segment of its work on fiber optics. According to an NSF assessment, the three centers' combined budget of $6.3 million included $3.6 million for photonics.[21]

The Origin of the Engineering Research Centers

The establishment of these programs in optoelectronics reflects the completion of a period of policy-making that began in 1983, during the Reagan administration. At a meeting early that year, Edward Knapp, then director of the NSF, and George Keyworth, the president's science advisor, found themselves in agreement that the foundation's engineering programs were inadequately meeting national needs. Keyworth asked Robert White, incoming head of the National Academy of Engineering, to advise NSF on the academy's views of the engineering program. With the assistance of a committee of notables from the engineering world, White prepared a paper, "Strengthening Engineering in the National Science Foundation," by July 1983.

Thereupon, the National Science Board issued a policy statement in August 1983 calling for an increased role for NSF in the engineering sciences. Having become aware that changes were brewing, the foundation's engineering staff also spent that summer producing a number of initiatives and position papers calling for alterations in the program. With Keyworth's approval, the NSF included in its October 27 budget submission to the Office of Management and Budget a request for $10 million to improve the quality of engineering education.[22]

It was two days later, in a meeting later seen as a milestone, that the contents of this initiative took clearer shape. Keyworth and other administration officials met with academic and industrial leaders to discuss computers in manufacturing. At that

October 29 meeting, George M. Low, president of Rensselaer Polytechnic Institute, described one of his institution's programs to encourage engineering students to work on joint projects in cross-disciplinary teams. According to an NSF staff member asked to write a chronology, this event galvanized the leaders of engineering. In time to meet the final January deadline for budget requests, administration officials decided that the budget for the new engineering initiative would be devoted to cross-disciplinary engineering research. In the meantime, NSF director Knapp again turned to the National Academy of Engineering for a description of what centers for cross-disciplinary research should look like.[23] Completed in February 1984, the academy's report, specified the centers' characteristics: purposes and size, annual funding, industrial participation, and means by which they were to be selected.[24] These recommendations finally shaped the research centers.

Standing as a prime motivator behind the establishment of the centers was a bill introduced by Congressman George Brown, Democrat from California. His bill would have removed the Engineering Directorate from the NSF and the National Bureau of Standards from the Department of Commerce, and combined them and other bureaus in a new National Technology Foundation. Though the bill never came to a vote, it revealed substantial doubts in Congress about the programs then in place for engineering.[25] Moreover, both Brown's bill and the Republican administration's counterproposal (for engineering research centers) expressed the widespread unease about competitiveness. Nearly all documents of the period reflect this.

An example is the cover letter with which Robert White, president of the National Academy of Engineering, officially submitted the academy's conception of the centers to NSF. The centers would have two related purposes, White wrote, of which the first was "to conduct cross-disciplinary research that would lead to the greater effectiveness and world competitiveness of US industrial companies." The second purpose was improved education of engineers.[26]

White's letter also raised a theme that was to become central in determining the form of the new centers:

> Since the major thrust of Engineering Research Centers is to improve the effectiveness and world competitiveness of U.S. industry, the industrial role in the centers' activities must be

prominent. The need for the engineering communities in both academia and industry to collaborate more closely is critical and overdue, and these Centers can be viewed as a significant step toward encouraging such collaboration.[27]

It is telling that this stress on collaboration in 1984 omitted reference to the same academy's report, of less than a year before, calling for NSF to set priorities for its engineering programs not just through collaboration but also through systematic reviews of engineering fields. The earlier report had called on the foundation to fund disciplinary reviews to provide background information "for priority-setting in the various fields of engineering research."[28] Of all the recommendations made by the academy's two reports on engineering research centers, this appears to have been the only one to have been ignored.

The establishment of the engineering research centers became the responsibility of Nam P. Suh, who was confirmed as the new NSF Assistant Director for Engineering in October 1984. The research centers were placed under a Division of Cross-Disciplinary Research. The division having been established, its staff had to oversee the decision-making procedure by which the centers would be chosen.

Selecting the Research Centers

Between 1985 and 1988, staff members processed 444 university proposals for the establishment of Engineering Research Centers. They designated eighteen centers, of which three would be substantially concerned with optoelectronics.[29] Like the individual projects funded by the Lightwave Technology Program, the centers were chosen through merit review, though of a more complex kind requiring several layers of approval.

In the first stage of the reviews, a panel of university and industry researchers who had shared knowledge of a broad technical area, such as manufacturing or biotechnology, culled through the proposals falling within their field. They evaluated proposals according to criteria that included potential research quality, contribution to international competitiveness, quality of advanced education to be offered, involvement by industry, and commitment by the university. A second-level panel, consisting of senior research managers and academic administrators, then reevaluated the recommended proposals, and visited the campuses.

According to a General Accounting Office investigation of the process, research quality was by far the most significant criterion in pruning each year's hundred or so proposals. Only after the second-tier panel returned from site visits, and narrowed its list to about a dozen finalists, was economic competitiveness intensively discussed. The panel recommended between four and nine proposals. The foundation's administrators then made final recommendations to the National Science Board, possibly applying secondary criteria, such as geographic balance and the avoidance of multiple grants to one institution.[30]

As the General Accounting Office investigators put it, since this is a program with the "ultimate goal of international competition," questions are raised on how the NSF chooses research areas that seem to be "relevant to the requirements of future competitiveness."[31] In two of its four program announcements, the NSF did mention favored technical areas. In 1988, for example, it gave design and manufacturing, advanced materials processing, resource recovery and utilization, and "emerging technologies," such as neuroengineering and lightwave technology, as examples of technological areas that would enhance international competitiveness. These were presented as representative areas, not as an exclusive list.[32]

While the NSF did attempt in two funding years to suggest broad technical areas for engineering centers, it did not study technical fields from the point of view of potential industrial effects. The NSF simply did not keep on staff analysts who could make these kinds of assessments. Asked about NSF's choice of topical areas for agency research centers, Lynn Preston, deputy director of the NSF division administering the centers, said NSF had stopped suggesting such areas. "Industry and PI's [principal investigators] have better knowledge of technical opportunities than do some bureaucrats in Washington." And the resouces were lacking at NSF to properly make such analyses. "This place is not a MITI," she adds, "and we should not try to make it into one until that becomes part of our mission. We shouldn't try to do it half-assedly."[33]

If NSF did not assess specific technologies such as photonics and their industrial effects, nor give priorities to technical areas, nor decide whether public involvement should be the same or should differ among technologies, such decisions did nevertheless fall to someone, somewhere. The decisions fell partly to the deliberations of engineering societies, and to the politicking and budgetary process, through which the policy was initially shaped. It fell to the review panelists, many from industrial firms, who chose the engineering centers. Most of all, it fell to the designated university centers themselves.

As clearly specified in the NSF's announcement giving information to universities applying for the program, the industrial purpose to which engineering centers respond was supposed to come not from the NSF but the applicants. The announcement listed as the first of an application's "key features" the following: "A clear and coherent vision guiding high quality fundamental research in areas critical to U.S. competitiveness."[34] The vision, then, was supposed to come from the interested parties, the universities themselves and (since private firms' commitments were required for complete proposals) from their interactions with industrial affiliates.

Research Centers and the States

The idea of using research centers as an economic development tool took hold of the states in the 1980s, such that by 1987 some twenty-nine of them supported such centers.[35] Many states established science and technology agencies, which had the selection and monitoring of research centers as their main function. At least six states set up centers in one or another subfield of photonics. Apparently without exception, however, the states took no further role in the new organizations, allowing them to set their own research priorities.

An example of one of these state agencies is Virginia's Center for Innovative Technology, created in 1984. The agency established five technology development centers, one a Fiber and Electro-Optics Research Center at Virginia Polytechnic Institute, to create "a critical mass for industrially oriented research."[36] Another example, New Jersey's Commission on Science and Technology, also established in 1984, funded several research centers at universities, two of them devoted to photonics. In Colorado, the Colorado Advanced Technology Institute supported an optoelectronics computing systems center at Boulder, helping it gain NSF recognition as a federally designated engineering research center.

New Jersey and New York were especially active in optics. In New Jersey, funds raised through bond issues in 1984 and 1988 paid for new construction as well as partial operating costs for the new centers.[37] The earlier bond issue established Rutgers

University's Center for Industrial Ceramics Research, which includes a large fiber optics materials research program. The 1988 Jobs, Education, and Competitiveness bond act yielded a $10 million endowment for an advanced technology center in photonics and opto-electronic materials at Princeton.[38]

An important founder of such research centers, the New York State Science and Technology Foundation, was older among state science agencies. Established by statute in 1963, it was originally dedicated to strengthening graduate science and engineering departments in the state and was motivated by the Nelson A. Rockefeller administration's view that the state was undergoing a brain drain. For the next eighteen years, the foundation occupied itself by quietly distributing grant funds to graduate programs in the state.[39]

Coming out of years of fiscal crisis and severe recession, the Hugh Carey administration's Economic Affairs Cabinet decided to include high technology in its 1978 election year agenda. The election campaign expressed the governor's interest in high technology matters. After Carey was reelected, a Governor's High Technology Council was set up to discuss the matter, several staff memoranda were drafted, expressions of support were made by high-level academic and industrial officials, and negotiations took place between the executive office and legislative leaders.[40]

The policy that emerged stressed, among other things, creating centers of specialized high-technology strength in universities around the state. The final legislation reconstituted the New York State Science and Technology Foundation and made the selection of Centers for Advanced Technology a main part of the foundation's broadened responsibility. Setting up the program in 1982, legislators found that

> Active collaboration between industry and the universities of the state in basic and applied research and development is essential if New York is to realize the potential for growth that new technologies represent. It is therefore the purpose of the state to foster such collaboration through the designation and support of several university-based, state-supported centers for advanced technology.[41]

In anticipation of the legislation, the Science and Technology Foundation commissioned a study by Battelle Columbus Laboratories, which found, mainly from a study of industrial census data, that the state's strengths could be described as being clustered in a number of technological areas, including automation, biotechnology, telecommunications, medical instruments, and optics.[42] Using this as a suggested list, the foundation sent out requests for proposals to the state's campuses in 1983 and received nineteen replies.

As did New Jersey's Commission on Science and Technology, the foundation processed the applications through peer review. The agency contracted with the National Research Council, which selected proposals and forwarded scores and comments to the foundation's board of directors. Composed of university, industry, and government officials, the board made the final selections. In the 1983 round, seven research centers were designated in as many academic institutions around the state.[43]

Two of the centers came to work in fields related to optics. One at Columbia University was to direct much of its research at optical communications. The other, at the University of Rochester, was designated the Center for Advanced Optical Technology. Each was awarded a round $1 million per year initially for five years, though that period was extended by legislation in 1987. Each was expected to have private firms match that amount and have them participate in the center's activities.

As in other states, the centers in New York were chosen partly to capitalize on regional technological specializations. In Florida, the state-supported Center for Lasers and Electro-Optics at the University of Central Florida at Orlando hoped to build associations with the state's aerospace firms, Martin Marietta, Harris, and United Technologies. The establishment of New Jersey's fiber-optics materials center at Rutgers and photonics center at Princeton reflected the proximity of AT&T's Bell Labs in Murray Hill, the regional phone companies' Bell Communications Research in Livingston, and the David Sarnoff Research Center (owned by General Electric) in Princeton. Rochester's Center for Advanced Optical Technology took advantage of a long local history of specialization in photographic optics at Eastman Kodak and Xerox.

In all states, the research centers were to play a part in interregional and interstate technological rivalry. As the annual report of the New York State Science and Technology Foundation explained, "A competition is underway among the states...to maximize their share of economic growth." "New York's key initiative in this contest," it continued, "is a program to encourage and support the development of technology-based industries."[44] One state's initiative to do so provided justifications for other states to follow suit. Explaining to a reporter New Jersey's decision to fund the Princeton photonics center, the vice chairman of the state's Commission on Science and Technology said that it was built because Florida, New Mexico, Arizona, New York, and California had their own programs in the subject.[45]

For purposes of interstate rivalry, a research center needed only to create a sense of regional technological identity. For such purposes, it had only a promotional role, and its substantive organization was irrelevant.

However, proponents who originally backed the new programs out of concerns about competitiveness, and a new breed of state professionals who administered such programs, also saw in the programs a more substantive purpose. As Vernon Ozarow, director of the Centers for Advanced Technology Program in New York State informally put it, the research centers represented a "lumpy" model of research support, as compared to the more distributed model of the earlier state and federal research grants.[46] By being designated as state centers of technological strength, the collaborative research organizations were to develop concentrations of technical skill that would develop the state or regional economy more effectively than would the diffuse distribution of grant funds. Potentially, a state or region could use such technological investment to develop selected industries.

How did the state science and technology agencies direct research centers toward such industrial and economic ends? New York State could not provide specific policy directions, since it did not employ persons devoted to assessing the effects of technological change on industries either in the Science and Technology Foundation or in any other agency. The state foundation could influence the centers' activities mainly by requiring the centers to orient themselves toward industrial concerns. The research centers were required to report on their relations with industry: industrial membership, patent declarations, visitors from industry, industrially oriented courses and seminars, and other indicators of collaboration.[47]

The New York State Science and Technology Foundation and other state science agencies, like the National Science Foundation, dealt with the new research centers mainly as auditors, keeping themselves at arm's length from the research agenda. Research priorities and their relationship to any industrial future were to be decided through collaborative interactions between academic researchers and industrial representatives.

ESTABLISHING COLLABORATIVE RESEARCH AT THE UNIVERSITY

Setting up the New Organization

Once state or federal requirements have been met, and a research center is established within a university, faculty researchers have relatively secure funding. The research center now gears up for operation. A chief is selected from senior faculty members and gains the title of "director," in addition to professor. An administrative assistant is hired to keep track of new bookkeeping, reporting, and public relations requirements. New letterhead, business cards, and brochures are printed, and a new sign goes up over the professor's door. In the way of life of the university, surprisingly little changes.

But the researchers now have to contend with the expectations of affiliated corporations. The corporations do not necessarily object to the professors' maintaining their prerogatives in directing the research, since many firms prefer to be arm's-length observers of promising research. What began as a policy for industrial resurgence becomes, within the hallways of academia, a funding mechanism.

When the University of Rochester received the New York State Science and Technology Foundation's designation as a Center for Advanced Optical Technology, it placed the center within the Institute of Optics, an organization already in existence for more than fifty years. Despite the name, this venerable organization in American optics is an academic line department within the university's college of engineering. The head of the institute in 1983, Kenneth J. Teegarden, gained the additional title of director of the Center for Advanced Optical Technology. The most substantial change that the new designation brought was the state's $1 million in annual funding, plus matching funds added by industrial participants, all available to be allocated at the discretion of the faculty. Indeed, the money helped the University of Rochester better pursue its missions—the institution could buy expensive laboratory facilities, expand the amount of research it conducted, and help support graduate students. The funds also helped support educational and research functions.[48]

But there was a catch. Matching funds had to be obtained through the establishment of a range of collaborative relationships with industrial representatives.

Recruiting the Industrial Affiliates

In the course of establishing a collaborative research center, the university faculty and administrators set out to find private companies that will affiliate with the center. To meet New York State's requirements for matching funds for the establishment of the Center for Advanced Optical Technology within the Institute of Optics in 1983, the University of Rochester turned to its long-time regional supporters, Eastman Kodak, Xerox, Bausch and Lomb, and Corning. All four joined, but Bausch and Lomb later dropped out when its optical instruments division was acquired by Cambridge Instruments, a British firm. As of 1989, the remaining three each paid upwards of $175 thousand per year to be an "Industrial Sponsor" of the center.

Similarly at Columbia University's Center for Telecommunications Research, such sponsors commited $200 thousand annually. As did the sponsors at the University of Rochester, these companies acquired the privilege of sending management representatives to meetings of a policy board and sending technical representatives to participate in a technical advisory council. The advisory boards met infrequently, once or twice a year, though the faculty interacted more frequently with corporate affiliates through symposia, retreats, phone conversations, and visits. The sponsors also gained royalty-free but nonexclusive rights to patents granted to the research center's innovations. And there were various programs of exchanges, workshops, seminars, short courses, newsletters, and advanced research reports.[49]

Like many other research centers around the country, the Center for Telecommunications Research maintained multiple tiers of affiliation. "Industrial Affiliates" paid $50 thousand for second-tier memberships, and "Industrial Associates" paid $10 thousand for the third tier. At research centers so structured, lower-level affiliates went to fewer workshops and meetings unless they paid additional fees, were not invited to advisory board sessions, and gained no rights or lesser rights to the research center's intellectual property.

Since industrial matching grants must be obtained to demonstrate to state and federal funding agencies that the research center interests industry, recruiting industrial affiliates is a critical early step in the life of every research center. In recruiting the affiliates, the founders of the center, especially those in engineering colleges, can build on a long history of informal relationships between faculty researchers and industrial counterparts. A professor of optical or electronic engineering typically has a range of relationships with corporate researchers in optics. He or she might have previously worked with them in an industrial laboratory, served as a consultant to their firms, studied with them in college, or had them as students. Professors with national reputations will have worked for years as consultants to corporations and defense agencies. Furthermore, academic and industrial researchers meet each other at the local chapter meetings and national conventions of the Optical Society of America, the International Society for Optical Engineering, and the Society for Imaging Science and Technology.

When the time comes to find industrial matching grants for a publicly funded research center, faculty members start with their professional acquaintances. This course of action is common to all research centers, not just those in optics. A survey of industrial participants in NSF-funded engineering research centers found that 66 percent of the respondents were personally acquainted with the center's faculty before it was established.[50]

It does not necessarily follow, however, that the companies provided additional funds to the university. Through the same period, the 1980s, when collaborative research centers were cropping up around the country, many corporations were reorganizing their philanthropic activities. For example, as a number of persons interviewed indicated in confidence, Eastman Kodak deemphasized the roles of corporate gifts committees awarding lump-sum all-purpose grants to universities. Funding decisions were turned over to operating units that would more carefully target their gifts in support of business strategies. In some cases, the company's funding for industry-university collaborative research in the 1980s represented the refitting of a traditional grant in a new guise to meet more carefully identified corporate needs as well as government's matching fund requirements.

In many firms sponsoring optics research, even those that shift gift giving to operating units, decisions on university funding are still treated as a philanthropic activity, albeit an activity with potential practical benefits for the firm. Treated as philanthropy, the grants qualify for favorable tax consideration. Corporate money for research centers is especially likely to be treated as philanthropy if it is not directed at a specific research project requested by the contributor but is, rather, pooled with the funds of other companies. "As soon as you talk about pooling research," says Robert R. Shannon at Arizona's Optical Sciences Center, "decisions get bumped up to a subcommittee of the board of directors, usually one that starts dealing with educational grants."[51]

Even though the policy makers who initiate the collaborative research programs expect the centers to have a role in long-term industrial strategies, the industrial participants often come to see their own role as an honorific one alongside academic colleagues, and as a philanthropic activity. To the extent that they take this view (and this extent varies considerably, as we will see below), the industrial affiliates are willing to give the senior university faculty the prerogative to use the pooled funds as they see fit.

Research Funding through Accumulated Designations

Further building on its research reputation and on the prestige of housing a New York State supported Center for Advanced Optical Technology, the University of Rochester's Institute of Optics acquired additional designations for its research. In 1986, through a $3 million grant handled by the Army Research Office, the institute was designated the University Research Initiative national center for optics.[52] In 1989, the Institute of Optics also came to house the new Center for Optics Manufacturing, another collaborative industry-university research program run jointly with the American Precision Optics Manufacturers Association and the Defense Department, and funded mainly by the latter.[53]

The practice of accumulating such designations and the money that go along with them is not at all restricted to the University of Rochester. At the University of Arizona, the Optical Sciences Center, an interdisciplinary program, brings together researchers from conventional science and engineering departments and houses the equipment and laboratories used by these researchers and graduate students in optics. In 1984, the center organized an Optical Circuitry Cooperative which received designation from NSF as an Industry-University Cooperative Research Center. A few years later, the same organization placed under its wing a new Center for Optical Data Storage mainly funded by industrial members.

At Columbia University's Center for Telecommunications Research, one of the three NSF engineering research centers that works in photonics, a more curious combination took place when essentially the same group of researchers succeeded in also winning designation as a New York State Science and Technology Foundation center for advanced technology. The state foundation was already naming a center for telecommunications technology at Polytechnic University in Brooklyn, so Columbia's center in Manhattan came to be known, for state purposes, as the Center for Computers and Information Systems. How do the state and federal designations relate to each other? The centers have different directors but share overlapping sets of faculty and students; "The relationship between the two is unclear to all."[54] But any confusion does not detain the faculty members or students. The complications show up only in the phenomenon of the professors proverbially wearing several hats, in the burdens of sending reports to the state and federal governments and undergoing audits, and in the elaborate bookkeeping.

A study of industry-university collaboration in California colleges has referred to such formally designated centers as "visible structures,"[55] and visibility is indeed one of their prized achievements. The government's designation of a research center brings prestige, a significant commodity at any university. More importantly the research centers bring money. As Robert Shannon, director of Arizona's Optical Sciences Center observes, "Lots of academic administrators see this as a new marketing strategy."[56] At a time of stringent federal budgets, when federal policy itself has become sympathetic to new research centers, such multiple designations become elaborate schemes for raising academic research and operating funds.

SETTING THE AGENDA IN UNIVERSITY PHOTONICS RESEARCH

The Pressures on the Research Agenda

By encouraging industry-university cooperation, the NSF and at least some of the states intended to remake the university research agenda to better prepare the region or nation for an era of technical rivalries. Where university centers for collaborative research in photonics were set up, how indeed did they respond to this broad public charge?

As a practical procedure, the answer is simple and similar among research centers. At the University of Colorado's NSF-funded engineering research center in optoelectronic computing systems, for example, the director sets the research agenda in consultation with heads of major research programs in the center and with the advice of members of a scientific advisory board composed of representatives of the center's industrial affiliates.[57] At the Rutgers University fiber-optics materials research program supported by the State of New Jersey, the director prepares a list of research options, accepts proposals from faculty researchers, consults with faculty colleagues, and takes suggestions from industrial advisors, all this leading up to a consensus on the distribution of research funds.[58]

But a listing of the steps can mask a complex and ambiguous dynamic. The participants in such an endeavor set the research agenda as a compromise between two diverging pressures, with a potential third having nearly no influence at all. The first is the desire of some corporate members to parcel out research projects, and the intellectual property that research occasionally yields, among corporate payers; faculty members go along in order to develop practical relationships with, and funding sources at, private companies. The second is industrial members' partially philanthropic attitude toward university research, combined with a desire to stay abreast of, but not directly involved in, technical change, something that can be accomplished by leaving the technical research decisions to the faculty's discretion; faculty members in turn naturally wish to have the agenda left to them so they can pursue research directions that promise academic reward. These pressures pulling at the research agenda leave open the question of a third one, raised by the policy-making that went into the creation of research centers, the desire to fit research choices into a broader technological and economic strategy.

The rest of this section elaborates this interpretation of the pressures tugging at the research agenda: demands to parcel out the resources, faculty and corporate intentions to maintain faculty control, and the policy intentions that originally led to the establishment of the centers.

Parceling the Pieces

In a few collaborative research centers, industrial participants have responded to the ambiguous privilege offered to them as advisors and partial funders of research projects by seeking to parcel out the projects among the paying sponsors. In the extreme case—no centers in photonics completely conform to this extreme—corporate laboratories in effect relegate research projects to academic faculty and graduate students. The corporations treat as a subsidy the tax breaks for corporate contributions, the potential for acquiring patents whose development was funded by federal funds, the academic institution's untaxed status as a nonprofit organization, and the direct government support for research centers.

The parceling out of research projects seems to occur more readily in technical colleges that have few or no advanced graduate students, a less well established tradition of scholarly publishing, and faculty instructors who depend on industrial contracts to supplement their income. A nearly pure example of parceling out can be found at Case Western Reserve University's Center for Applied Polymer Research. One of the NSF Industry-University Cooperative Research Centers, it is devoted marginally to optics, having dealt with the properties of optically clear materials, such as contact lenses, for medical uses.

As told in a historical narrative of the center, "Each of the research projects being pursued by [the center's] researchers is of primary interest to one and only one company, and there is relatively little transfer of industrial applications across the research agenda." Each company is given the right to patent inventions that emerge from the projects it sponsors. The affiliates do, however, set aside 15 percent of the center's budget for "blue sky" projects of joint interest.[59]

At Polytechnic University in New York City, the Institute of Imaging Science, which is supported primarily by industrial funds, encourages faculty members to write research proposals to the institute's member firms. Research funding depends on one-on-one relationships between faculty members and industrial sponsors.[60] The Rochester Institute of Technology, home of the Center for Imaging Science, maintains a for-profit subsidiary, RIT Research Corporation, which can link faculty researchers with external clients for proprietary research and consulting assignments.[61]

Universities with traditions of scholarly publication are less likely to promote direct faculty-business links but must still contend with corporate affiliates' demands. Until the 1980s, the University of Rochester's culture of academic scholarship saw little academic value in faculty members' seeking patents. Patenting is now positively encouraged. The university's Institute of Optics now allows private firms to sponsor specific research project with faculty members having compatible interests but, in keeping with university policy, requires that all research be made available for publication.[62]

At universities having a tenure track that rewards scholarly accomplishment and having faculty researchers with established reputations in their fields, the faculty members seem to have more leverage vis-à-vis an industrial sponsor who wants to parcel out the research center's resources. Such researchers prefer to direct their work toward projects that would bring recognition in their field's research community. Such aims mean that any short-term and parochial research interests on the part of industrial sponsors are to be discouraged. George Sigel of the Rutgers fiber optics program points out that such short-range demands are, as he puts it, the danger posed by collaborative research. "We shouldn't be an extension of an industrial lab," he says. "We shouldn't do tire-patching research, where industry comes in to say we want to fix this thing or that thing."[63]

Larger universities with doctoral programs resist pressures to parcel out research programs, but all research centers examined for this study do agree to convey, at least to the higher-tier industrial affiliates, partial rights to intellectual property developed by the research center. By a common formula, titular ownership of patents remains with the university. Affiliates acquire the right to review research results for potential patenting and to request delays of several months in the publication of results so they can decide whether or not to pursue a patent. Contracts sometimes also specify that the signatories will maintain confidences on corporate technical information.[64]

Such contractual arrangements received serious attention when research centers were being widely set up in the first part of the decade. For example, at organizational meetings for the Optical Circuitry Cooperative, a corporate membership group associated with the University of Arizona Optical Sciences Center, faculty and industrial advisers worked out a formal membership agreement in March 1984 that gave industrial members a royalty-free nonexclusive license to patents obtained from research sponsored by the group. When the members met again in January 1985, they found that a university administrator had altered this clause of the agreement to bring it into line with previous university policy. Several members of the group expressed such "extreme dissatisfaction" at this move that the university's chief research administrator agreed to restore the original language.[65]

Such concerns were typical at the early organizational stage of many research centers, but affiliates and faculty realized after some time that university patenting was not a very significant matter.[66] Universities just did not often seek or receive patents. If a faculty member's innovation indeed came to be patented, it was likely to be a device used in research and to have little commercial value. Of 38,127 patents granted by the U.S. Patent and Trademark Office in 1986, only 498, about 1 percent, went to universities. Of patents granted between 1963 and 1986, 6,807 were for lightwave technology. In a listing of organizational holders of these lightwave patents by the number of patents held, the academic institution having the most patents was MIT, which, with 31 patents in the field, ranked thirty-third. Even MIT, the most active patentor among universities, was awarded no more than three patents in any year in lightwave technology.[67]

The chance of gaining a commercial windfall from patenting is meager, particularly for a small and specialized research program. At the University of Rochester, which the patent office data show to be among the top twenty university patent holders in the country, the number of patents granted each year stayed in the single digits for all but one year between 1980 and 1987, and only a few of those patents would have been in optics.[68] (But in the course of the decade, as university administrators and faculty members gained experience in patenting and as obstacles to the patenting of biotechnology were overcome, universities in general may well have increased their success in obtaining patents.)

In the decade that university-industry research relationships multiplied around the country, industrial affiliates learned that they could sometimes redirect academic research programs toward subjects of immediate and proprietary interest in their firms, even though the results could rarely be appropriated in the form of licenses and patents. The extent to which academic research came to be oriented to the parochial needs of specific companies was not known. At least in photonics reseach in institutions with a record of research accomplishment, the tendency did not seem to dominate. Furthermore, many industrial affiliates, especially those representing large corporations that had a wide technical focus, discouraged the parceling out of research projects to short-term technical ends. They frequently preferred to let the faculty set the research agenda.

Leaving It to the Faculty

When a state or federal government designates a collaborative research center on a university campus, the faculty normally receive the event as an honor, a recognition of the importance of their institution and of their own abilities. At least in universities with a research tradition, the funds that go along with a research center allow the faculty to continue to pursue research directions that find respect among academic colleagues and provide doctoral research opportunities for advanced students. The state or federal grant is indeed worth celebrating, in the faculty's view, because it provides them with a resource to be used at their own discretion. Faculty members will then resist the parceling out of research projects for the narrow purposes of affiliated companies.

Corporate affiliates themselves frequently support the faculty's using the funds to set their own research plans. The corporate affiliates' willingness to leave the agenda to faculty is sometimes so pronounced that the research professors themselves are dismayed. A professor at the University of Rochester tells such a story of one of the Center for Advanced Optical Technology's sponsors. As he tells it, this company, "basically sends its check and does nothing else." Though it has sent an employee for advanced training in optics, the company "does not even take a hand in working out the research agenda of its own sponsored grad student."[69]

At Columbia's Telecommunications Research Center and other NSF engineering research centers, the members of the policy board, composed of management representatives from corporations, suggest "thrust areas" for research and "give us the larger picture" but meet only once a year, so "it is hard to get their time, and several can't make it to any one meeting."[70] At the NSF-funded University of Colorado Center for Optoelectronic Computing Systems, the director reports that individual sponsors, who meet twice a year, "don't say fund that or delete that. They feel they don't know enough about the technical details to do that."[71]

Several directors of research centers point out that industrial affiliates prefer to let faculty members pursue long-term projects in fundamental engineering problems. The companies want their staff to use the research center to stay aware of the front lines of technical research and keep track of talented graduate students who can be hired to bring up-to-the-minute knowledge into the firm.[72] Should a line of university research give rise to the promise of practical applications, the corporate research manager would rather follow through in an in-house laboratory and gain exclusive rights over the outcomes.

Industrial representatives, for the purposes of their participation on an advisory board, themselves often take on an attitude sympathetic to academic research. Those appointed to such positions have usually earned advanced technical degrees and have some familiarity with the way of life of the university. They may be under no duress to produce business results from the relationship, since their company's philanthropic wing is paying for the memberships. They may have strong colleagual ties with the faculty members and may consider themselves in an honorific position on the research center's board. They are then entirely willing to place the research agenda in the hands of the faculty.

Corporate representatives' unwillingness to take research centers seriously as business investments has dismayed some faculty members. As one observer put it, "The corporate members...do not know how to make the CAT [New York State Center for Advanced Technology] significant for themselves."[73]

A few commentators have recognized that corporate sponsors sometimes prefer to maintain their distance from university research that they fund. For example, William Hamilton observes that in research on radical new technologies, firms use external relationships, such as university research, as a "window strategy," to stay abreast of current technical developments.[74] A window on academic research implies that the companies looking through the window watch with interest for developments that come along. They sometimes want to influence the general directions of research but do not interfere in the details. This attitude complements the intentions of the faculty, who, by and large, want to maintain control over the research agenda.

The outcome is a compromise agenda formed, on the one hand, by corporate intentions to farm out research projects, and, on the other, by converging corporate-university interests in retaining faculty control. However any research center resolves the contrary pressures, it is likely to leave out of that final balance any attempt to look strategically at how its projects relate to the public purposes for which the center was created.

Whence Strategy?

Many of the early arguments in support of research centers stressed their potentially integrative character. They could provide economies of scale, critical mass, complementarities of skill, and other advantages that the simple distribution of funds to individual investigators might not itself achieve. Moreover, research administrators could apportion research projects according to an understanding of economic and technical ends. The research center could potentially apply strategic economic insight to the formulation of the research agenda.

Have collaborative research centers fulfilled this potential planning function? Neither New York nor most other states require their research centers to prepare strategic plans. At the New York State Science and Technology Foundation, applicants for designation as centers for advanced technology must prepare management plans, but no broader strategy. One state administrator, looking back at the formulation of the foundation's guidelines, explains, "We didn't have that much faith that universities could put something like that together."[75]

NSF guidelines for engineering research centers do, however, call for the amalgamation of research projects into an integral whole guided by an economic vision. According to NSF review guidelines for the evaluation of the centers, an "excellent effort" toward international competitiveness features a "coherent vision for the center, articulating its purpose, its rationale for a center configuration, and, importance for competitiveness." It also includes a "fully operational strategic plan," cross-disciplinary research such that the center can "contribute more than a collection of individual research projects," and research with a value over and above that of a collection of individual projects.[76]

In response to these requirements, the designated engineering research centers prepare strategic plans and periodically revise them in preparation for NSF reviews. The strategic plans of the Center for Telecommunications Research at Columbia University and the Center for Compound Semiconductor Electronics at the University of Illinois at Urbana-Champaign illustrate how the research centers have dealt with this requirement.

The Columbia center's strategic plan suggests a "vision of the future" consisting of "a globally omnipresent integrated telecommunications network operating at speeds of many billions of bits per second" and carrying combinations of voice, image, data, and graphic communications. To achieve this vision, the center "focuses on the goal of intelligent optical networks." Telecommunications must be all optical "because the problems of ultra-high speed transmission, switching, and signal processing demand that lightwave technology be pushed to its limit." The plan divides the center's efforts toward this end into core activities and "thrust areas" and describes the technical barriers and research projects in each area.[77]

At Urbana, the research center sets out to develop integrated optoelectronic circuits that overcome the limitations of conventional semiconductor microelectronics in which circuitry has become so miniaturized and dense that it has nearly reached the physical limits beyond which electronic signals interfere with each other. The authors of the strategic plan for the center divide the center's activities toward this end into three research thrusts: optical cross-bar switch, communications for optical interconnects, and integrated compound semiconductor devices. Under each division, the plan describes technical research objectives and technical problems to be solved. Under each, the plan also contains a section on industrial needs. In a section describing the research program for the optical cross-bar switch, for example, the "industrial needs" passage consists of one paragraph expressing industry's agreement that such technology would serve industrial needs.[78]

How do the centers formulate such a vision? One way that they do not is through economic or policy studies. None of the university research centers have tried to look more broadly at their technical area of inquiry through studies carried out by academic departments of public policy, social sciences, or business. The case is proven by the exception, which happened not in optics but in biotechnology at Cornell. Cornell's New York State-supported research center in agricultural biotechnology set up an occasional advisory board of agricultural economists from the university's agricultural college. A study they initiated predicted that bovine growth hormone would have dramatic, destabilizing effects on the dairy industry. "Interestingly," comments a publication of the Government-Industry-University Research Roundtable, "this is the only instance we have come across where policy analysis is integrated into a scientific program."[79]

According to NSF guidelines, consultation with industrial affiliates is supposed to give university researchers an understanding of industrial needs that can be incorporated into a strategy. It is this element that the NSF administration emphasizes when asked about such strategies. Strategic planning for technology may have been something that universities have not done well, but meetings with industrial affiliates on advisory boards and at special retreats has helped the faculty find out where their field is going and where the crucial technical problems are.[80] And research centers have recently been becoming better at producing strategic plans.[81]

Yet it remains unclear whether the companies' own strategies have played a role in the research centers' plans. Both at the one extreme of wanting to use the research center for cut-rate contract research and at the other extreme of considering it a window onto recent technical developments, industrial participants have no reason to share corporate strategies with academic researchers. Moreover, it is not at all evident that most corporate affiliates have a technology strategy, that they have appointed to the research center representatives who know what this might be, that strategies can be shared with academic researchers, or that strategies are compatible across companies. University research directors asked about their consultations with corporations all mentioned that they did consult in preparing strategic plans, but their contacts seemed to consist of sporadic conversations.

Asked how the strategic plan was prepared at Urbana, Joseph Verdeyen, former director of the research center, explained his involvement in one revision of the strategic plan: "It's a matter of looking downstream and looking at the serious problems facing the industry, and how this research will contribute. So we sit down and try to look at the future: What if we're successful? How could the results be used?"[82]

While envisioning the future, the research directors must also bear in mind the established technical interests and specializations of the faculty. At Columbia, when the administrative director was asked in 1986 how the research thrusts were chosen, he replied that "The original concepts go back to work done by professors even before the center was set up."[83] He amplified several years later that the center's research specializations evolved over the years and new researchers were hired to develop new specialties. Nevertheless, "all the research ultimately has to revolve around the expertise of the faculty. That's a primary intellectual foundation you work with."[84] Spokesmen for several research centers point out that the preparation of a research agenda must take into account the established specializations and skills of the faculty and the appropriateness of research projects for doctoral dissertation projects.[85] Drafters of a strategic plan must therefore be careful when setting it up to incorporate the tenured faculty's well-established interests and capabilities.

Though they do not rigorously investigate the economic implications of their technological field, research centers must nevertheless prepare strategic plans for the development of the field. They must do so while also keeping an eye to the unresolved tensions of parceling out projects versus deferring to the research interests of the faculty. One need not be of an especially skeptical turn of mind to think that the faculty members who prepare the plans must, paradoxically, be retrospective in their vision. They must make the plan respond to the interests of the collaborating parties, and primarily to those of the tenured faculty itself. Therefore, though research centers were originally seen as part of a new policy contributing to a more integral organization of industrially relevant research, the policy of delegating the planning to collaborative arrangements finally allows the research agenda to become the preserve of the interested parties themselves. Technological priorities emerge from a resolution of the agendas brought to the table by industrial affiliates and faculty members, and not from any coherent vision or strategy addressing technological change in the economy.

CONCLUSION

In setting up industry-university research centers around the country, state and federal policymakers so structured technology policy that allocative decisions having a public purpose were driven down to collaborative committees. Centers for university research in subjects of industrial relevance were selected through committees composed of industrial and university reviewers who chose centers mainly by criteria of technical quality. Once established, the research centers were expected to use their relationships with industrial affiliates to guide research in a direction that would support domestic industrial strength. While giving the centers such responsibility, the policies neglected to convey to them the resources for study, investigation, and deliberation by which they could make such planning decisions.

Since Washington and the state capitals distributed budgets to the designated centers more or less equally, the amounts could not be adjusted to any set of priorities representing the stated end of competitiveness. Within each research center, allocative decisions reflected the mixing of the faculty's and industrial affiliates' divergent purposes, purposes that had no clear relationship to the public ones for which the centers were established. Since the research centers had ambiguous missions, the universities with better reputations could take advantage of state and federal designations as a complicated funding mechanism. To the extent that industries allowed the faculty a free hand, funding decisions could be made through consensus of the faculty members themselves, without the need for broader peer review.

In accepting the idea that university-industry interactions in collaborative research centers can provide strategic guidance to university research, policymakers implicitly make assumptions whose truth has never been investigated. These include the following: that firms themselves have thought-out technology strategies, that managers who are aware of the strategy are the ones to represent the firm at the university, that managers attend meetings at the center and understand the purposes of the research center, that their strategies can be openly and honestly shared, that they are compatible across companies, that faculty members are prepared to heed industrial advice, and that public ends in research are identical to those of a group of industrial affiliates.

If interaction with affiliates is not a reliable source of strategic vision, then the availability of an intelligible source for such research allocation decisions comes into question. Not a single research center engages in studies of the economic reasons for pursuing a technology, of industries that will be affected, rival research abroad, or potential interactions with other lines of technical development. While the new, privatized technology policy compromises the traditional procedure of technical progress through individual grants competition and open academic research, it fails to convey to the new centers the defined mission and sources of knowledge through which they can contribute to the industrial aims for which they have been established.

SEVEN

Sheltering

THE MILITARY AGENCIES DISCOVER THE COMPETITIVENESS PROBLEM

Industrial Declines, Military Worries

U.S. industrial deterioration in the 1980s posed a problem not just for those concerned about prosperity but also for those who understood that military strength was bound to industrial capability. From the middle of the decade onwards, while one set of notables examined corporate-university relationships and industrial consortia, another set of blue-ribbon panelists looked at the relationships between military needs and American industry. Their message was that a more internationalized economy brought relative industrial declines in technologically intensive fields and that these declines posed a threat to military strategies founded on the assumption of this country's technological superiority.

As ever-more technologically sophisticated industries came to be located in Europe and East Asia, the U.S. Defense Department became more dependent on them as sources of advanced equipment. To an extent, foreign purchases simply reflected DoD's own efforts to reduce costs in response to budgetary constraints and the lower costs of imports. Moreover, as the decade progressed, foreign sources frequently could supply products superior to those domestically available. And as U.S. industrial capacity declined, the firms and trained personnel that might have produced advanced products entirely disappeared from the U.S.

In military circles, this situation was seen as impairing the nation's ability to mobilize: the foreign sources could be cut off in wartime. And even if some domestic capacity remained, many firms lacked equipment, trained employees, and large-scale facilities for the surge of production needed under wartime mobilization. It was such potential shortfalls in the capacity for industrial mobilization that most readily elicited military demands for policy toward selected industries.

But nearer the end of the decade, the relative industrial declines had come to seem more severe. A spate of studies suggested a much broader set of reasons for military concern about the economy. As would now be ever more widely recognized, the technological capacity to produce advanced weapons and command and control systems did not emanate just from internally funded laboratories. Only rarely and at great expense could the Defense Department bring forth a new technology entirely on its own. It depended on the full range of manpower and research capability in the private sector, not just that found among a select number of defense contractors.

In computers, communications technology, advanced materials, and photonics, the Defense Department relied on expertise in firms representing the breadth of the civilian economy. Deteriorating domestic capacity therefore threatened not just mobilization in wartime but the very capacity to maintain a technological lead over adversaries.

As panelists, consultants, and congressional hearings examined the issue, they did not fail to notice that the military establishment itself had resources that might serve well in attempts to revive American industries. It had rather large purchasing power, which could influence the technical directions of suppliers and contractors. It made large and risky technological investments, whose outcomes might give American firms advantages over foreign competitors. The defense budget had long included funds that succored and cultivated firms with which defense agencies had close contractual relationships. There was reason to believe that defense agencies themselves not only needed more capacious domestic industry but might have the wherewithal to bring it about.

Indeed, military industrial programs proliferated in the 1980s. However, the sum total of those programs did not necessarily grow greatly in their budgets. Traditionalists devoted to the truer pursuit of weapons systems and military readiness saw these programs as marginal at best. The military services' own technological plans paid scant heed to the industrial capacities that would be needed for the production of the more advanced systems they sought. Military industrial policy remained largely a clutter of discrete programs.

Only at the end of the decade, when the recommendations of the blue-ribbon committees were in, did the air begin to change. To leading segments of the military establishment, national security itself would depend on military ability to foster industrial strength, and this ability would rest on greater coordination, management, and planning. Amid the helter-skelter of disaggregative and collaborative responses, the massive U.S. military sector seemed to many a bastion of the coherence, purposefulness and—not least—money, that might revive American industrial standing. And the military establishment had unique political standing. Industrial programs ensconced within it would potentially satisfy corporate interests that were dismayed by the competitiveness problem, while also appeasing the complex of interests that support military appropriations. And, they could do so without raising ideological ire at the prospect of economic intervention. Since claims about the industrial value of military (and intelligence) programs could be used to shore up military budgets, this constellation of political forces could, ironically, even be strengthened by the end of the cold war and the fall of the Soviet Union.

To investigate these matters, one must engage on its own terms the elaborate ethos of technological deterrence and the byzantine bureaucracy that constitute the U.S. military establishment. One must peer inside the belly of the beast and, once again, photonic technology allows a particular perspective.

As we will see in this chapter, photonics in its several applications was inevitably high on the lists of those who thought about military technological readiness. Among those applications, image displays and optoelectronic circuits, and others, had considerable commercial appeal. But in these, Japanese and European researchers had sometimes achieved or were quickly approaching, technological superiority. U.S. military buyers might soon depend on industrial rivals for the most advanced photonic technologies. In anticipation of that development, photonics itself became one of the targets of military industrial policy in the 1980s.

After reviewing these applications, this chapter examines cases in military sectoral policy: the Strategic Defense Initiative's program in optical computing, DoD initiatives in optoelectronic materials, the military's role in setting up research on the manufacture of precision optical components, and the Defense Advanced Research Projects Agency's (DARPA) activities in display technology and other fields of photonics. DARPA held that if military agencies were to obtain the sophisticated technologies of the future at reasonable cost, civilian industry would have to develop the capacity to produce them for more than just military markets. Capitalizing on a mandate to pursue speculative next-generation military technologies, an implicit charter to make such decisions freely, and an ample budget and reputation, the agency could make sector-specific technological choices for industrial, as well as military effect.

These military industrial programs were poorly integrated with each other and with the services' own elaborate technological forecasting exercises. Though one discerns in them a recognition of technological and industrial interrelationships in the economy, the programs reflect a clutter of discrete choices. Only at the end of the decade, as military planners became more preoccupied with the civilian economy, did calls arise for more coordination, planning, and comprehensive management of military-industrial programs. The chapter's concluding section asks whether military-industrial initiatives are indeed industrial policy and how they represent a form of privatization. Before we turn to the description of military photonics, the rest of the present section considers the range of industrial programs in the defense agencies and describes the calls for building such programs into a more coherent military response to the plight of U.S. industry.

DoD Industrial Programs

The U.S. Defense Department and the armed services have long had a number of programs intended at affecting technological progress in industry. The oldest and most expensive of them is Independent Research and Development (IR&D).

With predecessors that date back to the 1940s, the IR&D program underwrites a substantial portion of the research costs of contractors doing business with the defense department. Contractors may use the funds for any research subject, as long as it has potential military relevance. According to reports, research managers prize the funds because they can be used to pursue technological avenues without the "micro-management" of legislative controls.[1] The program funded $1.9 billion of research in 1984.[2] Except in the sense that it supports military contractors only, the program is not sector-specific policy. Like the R & D tax incentive in the civilian world, it is a form of negative tax, transferring funds to private firms that make their own technological choices.

Among military programs directed at specific industries and technologies, the oldest is DoD's Manufacturing Technology program, commonly known as "Man Tech." Its purpose is to improve the productivity and technical responsiveness of manufacturers serving defense industrial needs. Man Tech funds the introduction of new manufacturing techniques and equipment into contractors' plants. The predecessors of the program can be traced to Air Force attempts in the 1950s to improve manufacturing methods in the aerospace industry. Later the Army and Navy started their own programs. As of 1986, each service operated its program independently, though with a small guidance team located in the Office of the Secretary of Defense. Man Tech funds stood in the range of $200 million in 1984, less than in previous years, with dwindling DoD program requests for the years to come.[3] Declining DoD requests for Man Tech have led to accusations that, because of tight budgets, DoD was sacrificing the development of long-term industrial capability in favor of weapons systems and other items more urgently desired by the services.

Since the late 1970s, the Defense Department has also undertaken a few highly focused research projects on exceptionally critical technologies in which the U.S. seemed to be losing its lead over other nations. As early as 1977, DoD internal reports found that the American lead in integrated circuit technology was being eroded. The resulting multiyear program in very high speed integrated circuits brought together teams from the three services under a joint program manager. The program's total cost came to $1 billion by the time of its completion in the mid '80s. Thereupon DoD established a new program under the same management to provide a still more advanced integrated chip, this one an optoelectronic chip built from gallium arsenide.[4]

Among military programs having to some extent ends in civilian industry, the one with the second largest annual budget (second only to the IR&D program) is DARPA. Operated independently of the services, DARPA has been seen, ever since the launching of Sputnik—the event that led to the establishment of the agency—as the site of the speculative research that would search for significant leaps of military knowhow. Since the agency deals with technical fields, such as computing, that are quite generic yet critical in military applications, repercussions for civilian industry are expected and indeed sought. By the late 1980s, as we shall see, DARPA took upon itself a leading role in generating civilian technologies that would strengthen industrial competitiveness for military advantage.

The year 1988 also saw the launching of Sematech, flagship among a fleet of military-led collaborative technological endeavors. Also known as the Semiconductor Manufacturing Technology Consortium, it was initially proposed by the Semiconductor Industries Association and funded by the Defense Department with industrial participation. It became the largest of the militarily led consortia, funded by DoD at $100 million a year. Smaller programs directed toward photonic technologies would inevitably cite Sematech (along with Japanese examples) as the model they wanted to follow.[5]

In the course of the 1980s, other programs were also set up with the intent or claim of producing spinoffs that would benefit American competitiveness. Most notable among them was the Strategic Defense Initiative. And there were several smaller programs, such as one by the logistical commanders of the services to examine industries whose capacity to respond under wartime mobilization was being lost because of the growing domestic dependence on imports. In all, by the late 1980s, the defense establishment had in place a set of disparate programs in which effects on civilian industrial competitiveness were variously intended, asserted, or actively pursued, often at great expense.

Reconceiving the Economy

Since these programs are already in existence, studies conducted during the latter part of the decade intend less to set up new programs than to better coordinate and manage the existing programs for mutual military and industrial benefit. Such studies confront the matter through an unusual economics, through vocabularies or conceptual schema that seek to eliminate the distinction between military and civilian sectors. Crucial terms in this vocabulary are *defense industrial base, defense technology base*, and *dual-use technology*.

To Robert B. Costello, under secretary of defense for acquisition, the entire North American industrial base is relevant to deterring aggression and winning wars. The industrial base provides—according to his examples—weapons systems from rifles to ships, components such as radars and engines, and logistical support that includes food, clothing, and medical supplies. And this base represents not just the capability to produce goods but also the technologies used in production.[6] "The defense industrial base," therefore, "generally comprises the same manufacturers that produce goods for the general public."[7] But the definition is too encompassing for practical purposes, so Costello's report examines the industries that account for 95 percent of DoD's purchases from the manufacturing sector. According to the results of the Defense Economic Impact Modeling System, 215 industries fall into this group, producing everything from steel springs to signs and advertisement displays.[8]

While, in this definition, the industrial base includes technological capacity, other reports break out the U.S. technology base as a subject of investigation on its own. After all, military preparedness does not rest just on goods, such as steel springs, but also on the technological capacity to prepare ever newer and more advanced weapons systems. According to the Office of Technology Assessment, the defense technology base

> is that combination of people, institutions, information, and skills that provides the technology used to develop and manufacture weapons and other defense systems. It rests on a dynamic, interactive network of laboratory facilities, commercial and defense industries, sub-tier component suppliers, venture capitalists, science and engineering professionals, communications systems, universities, data-resources, and design and manufacturing know-how.[9]

The report goes on to say that "the defense technology base is largely the same as the national technology base as a whole."[10] Military technological interests are indeed wide ranging, but some technologies still belong more or less exclusively to one sector. Supersonic transport is of a more military interest, while agricultural technology is more a civilian concern. Reports prepared late in the 1980s became concerned with a third class, dual-use technologies, "technologies that can have multiple, significant applications to military systems, and that can be employed extensively in civilian industry as well."[11] Despite the use of the word *dual*, the term tends to eliminate distinctions between military and civilian sectors, stressing the unity of technological development across sectors. It is these technologies that would be of particular concern in military planning. It is in these that the most is to be gained from more careful attention to the interrelationships between military and civilian investment.

The concepts of 'industrial base,' 'technology base,' and 'dual-use technology' seek to eliminate traditional conceptual barriers between the military and commercial sectors of the economy. They stress that substantially the same resource, the same base of knowledge and capability, serve both prosperity and security. This conceptual refinement can then serve as a prelude to proposals for the more careful management of shared military-civilian resources to benefit both military strength and industrial recovery.

Blue-Ribbon Panels and the Counsel of Coordination

Distinguished panels of experts have investigated the problems of the industrial base, especially from 1986 onwards. Representing the tenor of several of them, the 1988 Summer Study of the Defense Science Board, a senior advisory body to DoD headed by the president of Lockheed Corporation, declared as the first of its "principal findings" the following:

> Of greatest importance is the fact that the continued deterioration of the industrial and technology base diminishes the credibility of our deterrent. It is a national problem requiring a coordinated response by government and industry. If our nation is to ensure its security for the coming decade and beyond, it must adopt a strategy which links military strategy with a policy to ensure the availability of the industrial and technological resources on which operations plans rely.[12]

In the same vein, *Bolstering Defense Industrial Competitiveness*, a 1988 report of the Under Secretary of Defense for Acquisitions, observed as follows:

> Despite general economic prosperity, there is concern over America's ability to compete in the international marketplace. The Department of Defense also is becoming increasingly concerned. Many basic industries of importance to defense production have declined, threatening the responsiveness of our industrial base. Left unchecked, such erosion could rob the United States of industrial capabilities critical to national security.[13]

Even Frank Carlucci, the secretary of defense, wrote in his 1989 budget request that "Our economy and standard of living were once envied throughout the world, both fed by the marvelous machine we call our industrial base. But in the 1980s we are witnessing an erosion in this critical defense foundation." Dependence on offshore suppliers has increased and "our technological superiority has in some cases vanished." These are serious challenges, he continues. "Our strategy of deterrence and ultimately our national security depends upon the continued productive capacity of our industrial base."[14]

Other reports with similar messages were prepared by the Aerospace Education Foundation,[15] the Ad Hoc Industrial Advisory Committee to the Senate Armed Services Committee,[16] the U.S. General Accounting Office,[17] the Congressional Office of Technology Assessment[18], and a number of industry groups. And the recommendations that were made, though not always heeded, were also similar.

The reports stressed that defense agencies would have to deal more comprehensively and strategically with the industrial world. In the midst of blazing public rhetoric for free-market adjustment, the reports of the elite committees, even those of the Republican administration's military agencies, coolly addressed narrower audiences with prescriptions of coordination, management, strategy, and planning. Among the multitude of government studies that appeared on competitiveness problems, the military reports on the industrial base were among the very few in which the term "industrial policy" was unabashedly used.

Remarkable examples can be found in the 1988 Summer Study of the Defense Science Board. Besides the chairman, the president of Lockheed, members of the board included Norman Augustine, chairman of Martin Marietta; Malcolm Currie, chairman of Hughes Aircraft; Kent Kresa, president of Northrop; Mark Miller, president of Boeing; and a sprinkling of senior vice presidents, vice chairmen of the board, lieutenant generals, university presidents, and Nobel Prize winners. According to a DoD information sheet on the group, they constituted the "senior independent advisory board of the Department of Defense."[19]

Presumably not a naive group, they gave as the first of their recommendations the establishment of a cabinet-level mechanism to assess the nation's industrial and technological capabilities and formulate policies that would ensure that these capabilities accorded with national security objectives.[20] They drafted a proposed presidential directive to that end. According to the draft directive, the cabinet-level mechanism, known as the Industrial Policy Committee, would be chaired by the president's national security adviser and composed of representatives of federal departments. It would serve as a subcommittee of the president's Economic Policy Council. Nine of the committee's tasks are suggested in the report, including:

Review of major government policies and their impact on the domestic industrial and technology base. Review of government policies as they relate to globalization of the industrial base. Development of a plan for periodic industry-wide assessment of the rate of technology advancement and production capabilities compared to national security objectives. [And] development of policies throughout the government that foster industrial innovation, modernization, and productivity.[21]

The under secretary of defense for acquisition, in a report also released in 1988, similarly called for coordinated policies, though in less specific terms. Making no mention of a cabinet-level committee, it focused on what DoD needs to do in response to the internationalization of the economy:

the Department of Defense requires a synchronized capability to analyze the structure and performance of *critical industries*; the impacts on these industries of changes in economic policy, acquisition policy, and conditions of international trade; and the development and evaluation of policy instruments aimed at fostering a healthy defense industrial base and contributing to sustainment of a healthy national industrial base.[22]

And a few steps closer to the heart of the Republican administration, Frank Carlucci's budget report stated that recommendations (of DoD studies of the industrial base) were already being put into action. Steps being taken included "forging the right relations with industry" and "establishing defense industrial strategic plans that support our military strategic plans."[23]

DoD was by 1989 beginning to implement several programs for increasing analytical capability and improving coordination in defense industrial programs. These steps (we shall return to them later) fell far short of those recommended by the 1988 Summer Study, but they suggested a more activist DoD in the next decade's military policies toward industry.

Were the new ethic of DoD coordination, analysis, and planning in industrial matters indeed to be carried through, the agency's technical programs toward photonics would eventually be incorporated into a broader strategy. But through the 1980s, DoD efforts to strengthen industrial performance in photonic technology were, by and large, a clutter of discrete programs. All the same, military agencies devoted considerable attention to photonics and U.S. industrial capability in it.

PHOTONICS IN MILITARY USE

Photonics Among Critical Technologies

There is no better indicator of the military importance of photonic technology than the Defense Department's own list of critical technologies submitted for the first time in 1989 to the Senate Armed Services Committee. Entitled the "Critical Technologies Plan," its presentation to the committee was mandated by law as a way of forcing DoD (and the Department of Energy, which operates the largest government laboratories) to determine its technological priorities.[24] The plan listed and described the twenty-two highest priority technologies for the Defense Department's future (table 7-1). Of the twenty-two, arguably ten fall under photonics or rely substantially on it—they are shown in capital letters in table 7-1.

It is truly remarkable that, of the twenty-two technologies, the first fifteen have their use in computation, communications, imagery, signal analysis, and the suppression or disguising of signals. Only one item, air breathing propulsion, deals with a technology of locomotion. And only two, pulsed power and hypervelocity projectiles, refer to technologies of destruction.[25] The traditional paraphernalia of destruction—bombs, missiles, canons, and so forth—appear to have achieved destructive power that is quite adequate for military purposes. Bombs and planes have become lower-level technical problems. The true military problems are now ones of computation and intelligence: When and where to send weapons, how to decide where to send them, and how to prevent adversaries from knowing where to send theirs.

Hence, of the ten DoD critical technologies related to photonics, only one, pulsed power (needed for energizing laser weapons), concerns high-energy optics. The nine other technologies rely on low-energy photonics. Of these nine low-energy uses, a set of six (appearing as items 9 through 15 on the DoD list) encompasses a spectrum of applications in the sensing, interpretation, display, and suppression of signals and images. The other three consist of fiber optic communications (item 8), integrated optics (item 7) and compound optoelectronic semiconductors (item 2), whose technological importance is listed as being second only to microelectronics.

7-1
U.S. Department of Defense
Science and Technology Funding for Twenty-two Critical Technologies
Fiscal Year 1990

Critical Technology	DOD Funding in $ millions
1. Microelectronic circuits and their fabrication	$200
2. PREPARATION OF GALLIUM ARSENIDE AND OTHER COMPOUND SEMICONDUCTORS	100
3. Software producibility	70
4. Parallel computer architectures	80
5. Machine intelligence/robotics	70
6. Simulation and modeling	115
7. INTEGRATED OPTICS	25
8. FIBER OPTICS	20
9. SENSITIVE RADARS (including laser radars)	170
10. PASSIVE SENSORS	170
11. AUTOMATIC TARGET RECOGNITION (including signal processing)	75
12. PHASED ARRAYS	80
13. DATA FUSION (integrated sensors, surveillance, and analysis)	90
14. SIGNATURE CONTROL	*
15. Computational fluid dynamics	30
16. Air breathing propulsion	116
17. High power microwaves	50
18. PULSED POWER	65
19. Hypervelocity projectiles	100
20. Composite materials	110
21. Superconductivity	100
22. Biotechnology materials and processing	60
Total science and technology expenditure on critical technologies	$1,896
Total related to photonics	$ 655

* Not funded with DOD science and technology funding.

Note: Technologies closely related to photonics are in capitals.

Source: U.S. Department of Defense, *The Department of Defense Critical Technologies Plan* (Washington, D.C., 1989).

Electro-optics and Imaging

Contemporary U.S. military planning envisions an ever more sophisticated and more integrated set of instruments for sensing, recognizing, classifying, processing, and displaying information. Of particular value is information that can be captured and displayed in the form of images. Simultaneously, such planning seeks to suppress or disguise the information emitted from American bases, ships, and planes, to prevent enemies from similarly collecting information on our military operations.

Continuing military interest in these technologies is reflected in no less than six of the items on the 1989 DoD critical technologies list (table 7–1). Sensitive radars, passive sensors, and phased arrays seek better means of collecting information on the movements of adversaries. But the proliferation and added capacity of these sensors threatens to flood military analysts with oceans of data. Hence technologies of automatic target recognition are sought in the hope that they can automatically reduce the data to a volume manageable by human analysts. The automatic recognition systems would classify an object (say, a freighter versus a destroyer), recognize it (as friendly or hostile), and identify it (say, by a registration number).[26] The potential means include advanced signal processing technologies, including such exotic techniques as analog optical computing.

Through technologies of "data fusion" military planners hope to integrate information from various sensors and surveillance devices in a comprehensive analysis. A commander directing battle arrangements and a pilot making decisions in a cockpit would each similarly benefit from the integration of data and from the comprehensible presentation of data on a display device.

Techniques for sensing and analyzing data must be developed all the more because potential enemies are using advanced technologies to disguise the signals their own aircraft emit. At the same time, since adversaries are also developing advanced sensing devices, U.S. forces, too, must try to minimize the signals that adversaries can recognize. In defense terminology, the characteristics of an aircraft that make it recognizable through radar, the acoustic signals that make a ship detectable, and the visible profile of a tank each represent that vehicle's "signature." Technologies of signature control, appearing fourteenth on the list of critical technologies, reduce or disguise a craft's signature by changing materials, altering emitted signals, or camouflaging. (Alternatively, disguised signatures can make decoys and dummies seem larger or more menacing than they really are.) Stealth technology, which allows aircraft to become invisible to enemies, is a form of signature control. Optical communication and optical information processing are conducive to signature control, since they do not emit detectable radiation.

In the 1980s, such electro-optic technologies of detection, interpretation, imaging, and disguise were already being implemented in rudimentary form. In April 1986, when one hundred U.S. Air Force and Navy planes raided Tripoli and Benghazi, they used "advanced night vision systems" to fly at night at low altitudes, and targeted with "precision laser-guided weapons." An infrared imaging system kept track of the targets.[27] Precautions taken in Libya reflected the spread of the very kinds of technologies demonstrated by the raid. Adversaries also had abilities in surveillance and detection so, whenever possible, air raids and aerial combat had to take place beyond the line of sight. Pilots had to keep track of terrestrial and airborne targets, or threats, through imaging systems.[28] But in Libya and again in the 1990 U.S. incursion in Panama, many of the "smart" weapons went astray. Precision guided bombs and missiles were sometimes thrown off course by the pilots' swerving to avoid fire, by pilot error, or by clouds. A new generation of miniature circuits was, however, already increasing the accuracy of the weapons.[29]

After three decades of development, aided by technological progress in electronics and imaging, laser techniques that had once been solutions in search of a problem set out in 1991 in search of Iraqi military installations. As it seemed from the news reports from the Gulf, U.S. laser-guided bombs regularly made their way into Iraqi ventilation shafts and skylights, and cruise missiles navigated hundreds of miles until they either pinpointed targets or got themselves lost in the streets of Baghdad.

In the months after the Gulf War, newspaper and television coverage toned down the exuberant reception of precision weapons, noting that conventional bombing had turned out to be far more widespread in the war than the exotic warfare that received top billing. But such second thoughts were beside the point for planners of military technology. The carpet bombing of troops dug in under the sand and the destruction of Iraqi military traffic jams in the desert in daylight hardly required the expenditure of precisely guided missiles. Where they were used, satellite intelligence, reconnaissance management systems, antimissile missiles and their sophisticated signal-processing systems, and laser-guided bombs proved their efficacy in combat. The Gulf War, in the end, reinforced military technological directions that had already been prevalent.[30]

Fiber Optics

By the late 1980s, the various electro-optic systems of detection, surveillance, and imaging that were proliferating in planes were testing the capabilities of on-board computing systems. The Air Force and the Navy were letting out contracts to merge a plane's various functions in new integrated warning systems.[31] More processing capacity and speed would be needed, but weight could not increase much further. Optical communications on board fit the bill, while adding the bonus of being undisturbed by radiation and of emitting no radiation that could be detected by enemies.

Other technological developments sponsored by DoD also increased the need for such rapid telecommunications. After the successful completion of DoD research on very high speed integrated circuits in the early 1980s, the introduction of such circuits further sparked the desire to link more potent computers with high-speed telecommunications.[32]

The requirements assuring the security of military communications also played a role, since encryption schemes required that voice communication be transmitted as digital signals. Such digital encryption consumed ten times as much communications capacity as did conventional voice lines, making the high capacity of fiber optics particularly desirable. Since fiber lines cannot be tapped without interrupting the light signal, the new technology also served security of communications.[33] Unlike electrical signals, light signals do not radiate, so fiber optics allowed the armed services to conform to top secret National Security Agency regulations meant to prevent military communications from emitting detectable radiation. Fiber optics eliminated the fire hazards caused by electrical components on aircraft. Optical signals were also immune to the electromagnetic pulse that would interrupt or ruin electrical communications after the detonation of a nuclear weapon in the atmosphere.

The light weight and flexibility of the fibers suited them for mobile applications in a battlefield. And research on fibers that would retain their signals over oceanic distances yielded a Navy plan to line the ocean floor with a grid of fiber-optic lines for tracking Soviet submarines.[34] For all these military uses of fiber optics, it seemed by 1989 that the technology would eventually have its most profound military effect not directly in the form of communications lines. Rather, the proven worth of fiber optics, and the massive communications capacity it made possible, would play a pivotal role in decisions to pursue the most prodigious of calculating weapons, the optical computer.[35]

Optoelectronic Circuitry and Optical Computing

In the DoD Critical Technologies Plan, compound semiconductor materials, used for producing a sandwich of materials with optical and electronic properties, have an importance second only to that of silicon microelectronics. DoD develops compound semiconductor technology, particularly gallium arsenide chips,[36] because integrated circuits built with them are many times faster than their silicon counterparts. Being faster and more resistant to radiation than silicon circuits (but not as fast or as resistant as all-optical circuitry could be), compound semiconductor materials are, according to the DoD document, "critical building blocks for DoD electronic warfare, radar, smart weapons, and communications systems."[37]

Several working prototypes of gallium arsenide chips were already in existence by the late 1980s. What remained most troublesome was fabrication technology. The production of the semiconductor wafers was still slow and painstaking, falling far short of the mass-production techniques available for silicon semiconductor chips.

Integrated optical circuitry, which the DoD refers to restrictively as "photonics," appears seventh on the list of DoD critical technologies. DoD work on such photonic devices encompasses optical memories, optical signal processing, optical circuitry, and integrated optoelectronic networks. If the means can be found to integrate these optical functions with minimal use of electronics, the resulting optical information systems will far exceed the speeds of contemporary computers and eliminate optoelectronic connectors that presently undermine reliability.

Some forms of optical processing were already in use in 1989. According to the DoD technologies plan, integrated optical processors were already in use in "front end applications" such as sensors. In electronic warfare and undersea surveillance such sensors are already producing data at rates "surpassing the capabilities of electronic processors."[38] Rudimentary optical processing systems, along with the proliferation of optical fibers and electro-optical sensing and imaging, themselves create the demand for further integration through optics. Moreover—as we shall see in the next section—in the proposed design of the Strategic Defense Initiative, for a five-minute decision on a missile attack to be at all possible, the defense system requires computer processing so rapid that only integrated optics will fit the bill.

By the end of the 1980s, researchers realized that optical computing was a long-term technological gamble, the success of which depended on several breakthroughs yet unachieved. DoD maintained integrated optics among its most critical technologies, but predicted that optical integration would evolve over a period of twenty years.[39] By contrast, compound semiconductors such as those made of gallium arsenide offered the prospect of short-run technical feasibility.

Military Photonics and Industrial Capability

As the Critical Technologies Plan itself recognized, military agencies could not possibly invest enough to ensure an American technological lead in every field of photonics. In these waning years of the Soviet Union, the Soviets were not the primary problem. The plan said at several points that they were reassuringly behind in most applications of photonics (though the plan did not fail to mention that Soviets were, as ever, catching up). The greater technological challenge arose from foreign research that sought a commercial payoff from photonics. To the extent that photonic technologies had commercial potential, DoD had to depend on U.S. industry to complement and reinforce DoD research.

But such industrial reinforcement was not forthcoming. As the Critical Technologies Plan showed, but did not say in so many words, this country led other nations in the very photonic applications—pulsed power, phased arrays, passive arrays, laser radar—that had the least industrial value.[40] In imaging, integrated optics, and compound semiconductors, which were of potentially enormous commercial value, U.S. technological standing was much weaker. In the commercial application of display technology, for example, Europe and Japan appeared to be forging ahead of the United States.

Also, according to the Critical Technologies Plan, "Japan leads the world in converting fiber optic technologies to various commercial applications"[41] In integrated optics, which will have revolutionary effects on future computing and communications, the U.S. seems to be leading, but Japan "is pursuing research and development in all areas of optical processing" and "has a large, well coordinated program addressing all aspects of optics."[42] And "Japan is the undisputed leader" in gallium arsenide materials technology, which is becoming an ever stronger competitor to silicon microelectronics.[43]

Below the waves stirred up by Desert Storm, the American reliance on Japansese technology for much of the war's most advanced technical devices remained only a quiet undercurrent. A journalist reporting on the subject found it very difficult to come by information on the use of Japanese products in U.S. weapons. Pacifist sentiments and Middle East trade ties in one country and sentiments of recent glory in the other made both countries uncomfortable about the matter. The journalist concluded that

> For the U.S., the superiority of its high-tech weapons in Operation Desert Storm has served to emphasize an increasing dependency on Japan's electronics industry, including semiconductor casings, flat video screens and imaging devices.[44]

Military Industrial Programs in Photonics

In retrospect, the Gulf War highlighted what U.S. military officials had been learning for a decade: in those photonic technologies that had significant commercial applications, technological strength was shifting overseas. When the only rival in such fields was the USSR, the Defense Department could regain technical advantage by reshuffling its technological investments. But when other advanced capitalist nations began to surpass the United States in some fields of photonic technology, DoD realized it must reassert domestic technological strength through effects on a number of domestic industries, and not just those reliant on DoD contracts.

As we shall see in the forthcoming sections, military programs asserting an industrial purpose were underway in optical computing, optoelectronic circuitry, precise optical components, and high-definition displays. The new policy initiatives were necessarily sectoral ones, since they had to concentrate on technologies that DoD found critical. And their aim was not just to have these industries produce defense items, since that could be accomplished simply through traditional contracting and acquisitions. Military industrial policies now had to build the abilities of American industry to survive in international trade.

INITIATIVES IN OPTICAL COMPUTING AND OPTOELECTRONIC MATERIALS

In the mid-1980s, the Strategic Defense Initiative was plausibly presented as the source of future U.S. industrial leadership in exotic new technologies. By SDI's own representations, optical computing would be among the foremost of the program's contributions to American technological welfare. But SDI was under pressure to demonstrate the practicability of the peace shield. When it turned out that outcomes from optical computing would exceed several budgetary cycles, SDI turned away from optical computing.

It seems in retrospect that protestations about industrial effects emerged more from SDI public relations than from concern about the domestic computer industry. Since other military research offices also desire to bolster budgets through claims that their research has industrial benefit, such claims in a military program cannot be taken at face value as indications of sectoral policy. SDI's investment in optical computing exemplifies the kinds of ambiguities we face in assessing military industrial policy.

Compared to the promise of all-optical computing, the projected advantages of hybrid optoelectronic circuits were more modest, but much more achievable. By the early 1980s, gallium arsenide wafers were available for missile guidance systems, radar apparatus, equipment used for electronic countermeasures, and communications systems requiring high-performance circuitry that was hardened against electromagnetic interference. But if the optoelectronic wafers were only to be manufactured for military needs, they would be inordinately expensive. The development of the wafers for civilian applications took place largely in Japan. The DoD therefore undertook initiatives to improve industrial fabrication techniques for gallium arsenide devices. It also seems to have tried to build a domestic optoelectronic industry directed at commercial applications. But, as we shall see below, one initiative toward such ends, a collaborative optoelectronic materials research center, came into being not through a concerted military technology policy, but through the conjunction of unrelated political events. Since the optoelectronics program was to take the form of a collaborative industry-university research center—albeit one under military leadership—it would not be easily distinguished from the collaborative form of privatized industrial policy.

SDI and Optics

Amid the controversy over SDI, spinoffs for industry and science became one of SDI's significant selling points. Even those who doubted that SDI would ever shield the skies suggested that its technological consequences for industry might indeed be real.[45]

As did other aspects of SDI, its potential for civilian technological spinoffs generated rancorous disagreement. SDI might well benefit industry, opponents said, but the cost would be exorbitant. And in hiring scientists and engineers for SDI research, the program would remove them from commercial research and unclassified academic work, thereby inhibiting conventional research. Yet the technical investment was large enough, and technical fields exotic enough, that SDI was seen by some as a form of U.S. industrial policy.[46]

If SDI was going to have significant effects on any technological field, it would likely be photonics. According to James A. Ionson, director of SDI's Innovative Science and Technology Directorate (the SDI wing responsible for speculative technological investments), programs in sensors, lasers, and optics formed the bulk of the directorate's programs. "If you add them all up," he said, "electro-optics comprises a good 60 percent of the overall [Innovative Science and Technology] program."[47] At least through 1986, a significant portion of that money went for one particularly speculative photonic technology, optical computing. The *SDI Monitor*, a newsletter that follows Star Wars contracting, stated that Ionson's directorate spent $18.5 million on optical computing (apparently in 1986), accounting for most of the nation's expenditure on optical computing research.[48]

At least in the early years of SDI, the technical appeal of optical computing was indeed irresistible. The program entailed bizarre technical demands for simultaneously coordinating arrays of orbiting sensors, identifying nuclear missile launches and tracking the missiles, distinguishing actual missiles from dummies, properly positioning orbiting mirrors, directing laser weapons or kinetic energy weapons at the missiles, and determining which missiles had been hit.

These items of space defense had to be accomplished by equipment that would not be ruined by the electromagnetic pulse that would follow a nuclear explosion. All the logic favored light-speed computing.[49] The logic was illustrated by William J. Broad, who in mid-decade interviewed weapons researchers on the technical requirements of an adequate Star Wars system. He quoted one researcher as follows on the need for optical computers:

They'd also be faster than regular computers... During an attack you've got about five minutes for decisions and battle management. It doesn't matter if your computer program is cost effective. What matters is if you've got the computational ability to predict the trajectories fast enough to kill the warheads before they kill you. It is absolutely essential to get that speed.[50]

It is properties of light that will conceivably allow optical computers to operate so fast. Since numerous light rays can pass through a lens, an optical computer could in theory accept multiple inputs at the same time; so it would be characterized by "massive parallelism," while most electronic computers can only accept data single file through a main processor. And since light rays, unlike electronic signals, do not appreciably interfere with each other, optical circuitry could be far denser than that found in electronic computers.[51]

But some specialists had grave doubts about the feasibility of optical computers. Optical computing, if it were to work, would have to keep signals from degenerating over long calculations. It would have to operate without emitting too much heat. And, not least, efficient counterparts to silicon would have to be found for switching the light signals efficiently. In some minds, optical computing was "wishful thinking on the order of: if we had some ham we could have some ham and eggs—if we had some eggs."[52]

In the face of these technical obstacles, only one venture firm appeared in the field through 1986, and it folded quickly.[53] AT&T was actively interested in optical computing, despite the risks. If means could be found not just to transmit light signals along fibers but also to process them optically, the gains would be enormous. Alan Huang, AT&T team leader in optical computing research, even suggested that AT&T "is going to become the photon company."[54] But in 1986 IBM remained the home either of realists or silicon stalwarts. A spokesman declared that "nobody at IBM is either working on optical computing or thinks it's a good idea."[55]

Had there been significant interest among private firms, SDI might have welcomed the added domestic technical capacity and even endeavored to build upon it. But computer firms balked at the technical challenges. (AT&T and other telecommunications firms had incentives strong enough to persuade them to invest in the field despite the risks. Links with defense agencies, if there were any, were not documented.) SDI and other military agencies would have to pursue optical computing largely at their own expense.

The dearth of industrial interest did not, however, prevent SDI from asserting that it's interest in optical computing would have civilian benefits. The SDI organization could give some evidence for the claim because in 1985 it established a military-university-industry collaborative research group dedicated to optical computing. Budgeted by SDI with $9 million for three years, it was headquartered at the University of Alabama at Huntsville. The funds led several universities to join with Huntsville, including Carnegie-Mellon, MIT's Lincoln Lab, and Harvard. Private firms, such as BDM Corp., also joined, though several of them were better known for military contracting than for a substantial interest in civilian computer industries.[56] H. John Caulfield, professor at the University of Alabama and head of the Huntsville consortium, said in 1985, "My instructions are to aim at a significant breakthrough."[57] He planned to emphazise research at universities and small companies. And, at least to Caulfield, Japanese competition was as much a worry as strategic defense.[58]

Spinoff PR at SDI

In 1986, at the height of the Star Wars controversy, as SDI's technical programs were getting underway, so were initiatives in the realm of public relations. Judging from the press coverage, one can surmise that SDI spokesmen and laboratories became readily available to reporters who wanted to write about civilian spinoffs from the program. Numerous articles appeared in the popular press on the subject. A representative example appeared in the *New York Times Magazine*, whose 1986 article was subtitled "The controversial defense system is yielding technologies that seem sure to change the world."[59]

The author of the article cited "experts" that SDI's computer research "is helping to bring into being powerful tools whose civilian counterparts will have incalculable scientific value."[60] Also in 1986, *Business Week* quoted SDI's program manager for battle management systems on the prediction that optical computing "could revolutionize the way we think about computing, much as the semiconductor chip revolutionized electronics 30 years ago."[61] Ionson, of SDI's Innovative Science and Technology Directorate, was more exuberant. Comparing optical computers to the supercomputers produced by Cray Reseach, he predicted that "ten years hence an optical computer packing the power of 2,000 Cray X-MP's will fit in a suitcase."[62]

As early as 1986, Caulfield, head of the Huntsville program, openly cautioned his colleagues about premature expectations for optical computing. He warned them not to repeat the mistakes made during earlier enthusiasm about holography. In holography "we let our justifiable enthusiasm cause us to underestimate future difficulties. In short, we promised more than we could deliver." Despite ambitious claims, he continued, optical computing might only be where electronic computers were in the 1940s.[63]

Technical results indeed emerged very slowly. SDI allowed its contract with the Huntsville joint research group to lapse. If Caulfield's prediction was right that optical computing stood where electronic computing did in the 1940s, then decades remained until research yielded practical results. But the politics of Star Wars necessitated a far faster time scale, and SDI support for optical computing waned. In the political maelstrom, SDI's claim of industrial spinoffs had surely emerged more from worries about budgets and public support than about competitiveness.

DoD and Optoelectronic Materials

Compared to the initiative in optical computing, the investment in optoelectronic circuitry represented a more readily achievable technical ambition. Optoelectronic integrated circuits made from gallium arsenide were demonstrated occasionally in the late 1970s. The 1980s brought a rapid succession of more advanced circuits.[64]

Unhappily for defense plans, however, most of the advances of the late 1980s occurred in Japan. If this critical technology was to be available to DoD, the agency might well have to obtain them from Matsushita, Fujitsu, or Nippon Electric.[65] Otherwise, DoD could hope to engender domestic suppliers through the traditional tools of DoD acquisitions policy. But in the modest quantities that would be needed, the unit costs for the optoelectronic chips would be prohibitively high, even by DoD standards. It made sense for DoD not just to attend narrowly to its own acquisitions needs but to build an optoelectronic chip industry that also served a civilian market.

DoD undertook programs to do so in the 1980s. As in other military research programs, attributions of credit and responsibility in the bureaucracy seem convoluted to an outsider. The organization—such as SDI, DARPA, or the Air Force Office of Scientific Research—which funds the projects, might be far removed from the bureaucratic locale that performs it, though both will likely take the credit. And much is also classified. From what can be ascertained, DARPA began funding pilot production lines in compound optoelectronic circuitry in 1982.[66] Whether in a related or separate program, the Navy's Man Tech program sponsored research on the industrial fabrication of gallium arsenide in 1985.[67]

In 1986–87, upon the completion of the DoD research on high-speed integrated circuits, the DoD office that coordinated the program was given responsibility for a new multiyear effort to produce optoelectronic chips. Referred to as the Microwave, Millimeter Wave, and Monolithic Integrated Circuits Program (the MIMIC program), it was intended to coordinate advanced research on gallium arsenide chips.[68] By the time that pilot production was expected to begin in the late 1980s, the program's total cost was to be in the hundreds of millions.[69]

In 1987, the Defense Science Board reported that DoD demand for advanced chips was still too low for reliable and cost-effective production. Since the DoD had invested "hundreds of millions" in developing lines of production, the Science Board suggested, the problem deserved serious attention. Several alternatives for further action were available. They ranged from "subsidizing early pilot plant production to stimulating use of the lines for commercial components."[70] One alternative, that of subsidizing pilot production, was indeed implemented in 1988. The MIMIC program funded four separate corporate joint ventures for the production of optoelectronic components.[71]

Another alternative, stimulating optoelectronic production for commercial purposes, was more troublesome. "We would like to have a gallium-arsenide industry in the U.S.," a top Pentagon official said in 1988. "But having said that, how do you do it?"[72]

Centers for Strategic Optoelectronic Materials

In keeping with the emerging traditions of the 1980s, one way to do it was through military-led university-industry collaborative research centers conducting research on strategic optoelectronic materials. The initiative emerged, however, not through concerted DoD industrial policy, but through rather circuitous politics.

The initiative took shape in 1985 in contacts between DARPA and the Institute of Electrical and Electronics Engineers (IEEE).

Within IEEE, the proposal for such an initiative was drafted by the Task Force on Defense Electronic Materials, composed of more than a dozen chief executives and corporate research directors in materials science, as well as a DARPA representative and academic members.

The task force suggested in a report that, because of the growing importance of high technology in defense, the meaning of "strategic materials" had changed. Formerly, such materials had consisted of naturally rare substances, such as chromium and manganese, which were in short supply in the U.S. In the present technological world, however, the term should include materials such as gallium arsenide, production of which depends on highly specialized facilities and skills. Shortages of such advanced materials would be as severely limiting in a military emergency as the absence of chromium or manganese.[73]

As the initiative evolved, DARPA and IEEE urged that "the need exists for an innovative R & D management effort to accelerate the pace of developing these critical technologies and rapidly transfer that technology to the domestic producers of semiconductor materials, devices and manufacturing equipment."[74] They recommended the establishment of a research center in one particular set of strategic materials, compound optoelectronic semiconductor materials. It would be funded through DARPA; governed jointly by government, industry, and university participants; and operated as a nonprofit organization. It would cost around $125 million.[75]

But such funds were not easily acquired in the late 1980s. The proposal languished until, as we might recall from chapter 5, Pete V. Domenici, senator from New Mexico, attempted to earmark a congressional appropriation for an optoelectronics center at the University of New Mexico. Twelve and a half million dollars (one tenth of the amount that IEEE had sought) for the center was indeed authorized in 1989, but the wording of the authorization was so vague that it was unclear that the money would have to be spent in New Mexico. Reputedly dismayed at the growth of academic pork barrel appropriations, Sam Nunn, chairman of the Senate Armed Services Committee, used his influence to make sure that the funds were to be disbursed through open competition among universities.

DARPA was given responsibility for handling the funds. And DARPA promptly used the prerogative to serve the intent of the old IEEE initiative by soliciting proposals for the establishment of the optoelectronics materials center. With only $12.5 million, it was initially expected to be a much more modest program than the proponents had originally hoped. But DARPA administrators were so favorably impressed by the quality and variety of the proposals submitted that they decided to appropriate internal funds in support of two additional centers for optoelectronic research.[76] The Optoelectronic Materials Center came to consist of a consortium of the University of New Mexico, Stanford, California Institute of Technology, Sandia National Labs, and MIT's Lincoln Laboratories. The two other centers, a National Center for Integrated Photonic Technology and an Optoelectronic Technology Center, each similarly comprised a number of universities and laboratories. The centers for optoelectronics came to be funded in part through purposeful Defense Department industrial strategy in which DARPA played the leading role and in part through a fortuitous convergence of political events.

SDI as Industrial Policy?

SDI's efforts in optical computing and DoD programs in optoelectronics serve to illustrate pitfalls of any sweeping attempt to characterize military industrial policy. In SDI, as in other military agencies, claims that military programs have industrial effect cannot be taken at face value. They might well represent public relations rather than industrial policy.

In DoD's program to improve the fabrication of optoelectronic materials, the motives were lowered costs of acquisition and domestic self-sufficiency. The development of a domestic optoelectronic chip industry that produced for commercial as well as military markets would serve the purpose. But decisions to go forward on a joint military-industry-university research program on optoelectronic materials appear to have been made in some of its stages more through happenstance than concerted strategy.

In that the programs try to shape the conditions of industrial technology, DoD's initiatives in optoelectronics do reveal the operations of sectoral policy. Signs of purposeful military planning for industrial effect become clearer in the cases discussed below.

PRECISION OPTICS AND MILITARY MOBILIZATION

In 1986 and 1987 several teams in the defense agencies undertook or contracted for studies of U.S. industries critical to national security. Among these studies, one commissioned by the Joint

Logistical Commanders examined the capabilities of the precision optics industry under military mobilization. The Logistical Commanders' study, along with no small measure of lobbying by the industry, led by 1989 to the establishment of a research center to develop means of automating the manufacture of precision optics.[77]

"Precision optics" refers to the lenses, prisms, mirrors and other optical elements used in technically demanding applications, such as lasers and sensors, as well as in high-quality optical instruments, such as military binoculars. According to Robert Leshne, head of the American Precision Optics Manufacturers Association, precision optical components are used in missile guidance and control systems, smart munitions, fusing systems, fire control systems, optical communications, and devices that resist electromagnetic interference, among other applications.[78]

The U.S. military agencies constituted the single largest market for the world's precision optics. American producers were quite effective, at serving this market, but by 1987 they had no share at all of the civilian market for optical components.[79] Moreover, the foreign producers, who fully dominated commercial markets, were encroaching on the U.S. military domain as well, raising fears of domestic military dependence on foreign suppliers.

The U.S. response took the form of the Center for Optics Manufacturing at the University of Rochester. The center had as its purpose "to strengthen the optics industry base in the United States by developing new manufacturing technologies."[80] To that end, it expected to have research programs in computer-aided design and manufacturing methods in precision optics, and to organize on-the-job training programs for the industry. Though headquartered in Rochester, it planned to establish branches in other university optics programs in the United States and in two-year colleges involved in technical training.

The FAR Study

Military concern about precision optics revolved primarily around the ability of U.S. producers to meet military needs during the surge of production expected in the outbreak of war. In 1985, a study by the Army Materiel Command, following upon an earlier study by an ad hoc study group, found that the domestic optical industry would be unable to respond in such circumstances, especially if declines continued. The study recommended the implementation of the Federal Acquisition Regulation (FAR), which requires DoD to restrict purchases of foreign optical materials in order to preserve the domestic industry.

Since the study dealt with the needs of only one service, defense officials organized still another study, this one conducted by the Joint Group on the Industrial Base, a subdivision of a committee of logistical commanders from the services. The Joint Group appointed a precision optics technical committee, which joined with the U.S. Department of Commerce to conduct a study of U.S. producers of optical materials and optical components. Under the Defense Production Act of 1950, the joint group had the authority to require firms to respond to a questionnaire and allow official visits to their plants.[81]

Their study, completed in 1987, found that imports accounted for 98 percent of the total U.S. consumption of optical elements, but this measure included the vast quantities of low-quality lenses, such as the ones used in inexpensive cameras. Data on the dollar value of imports reflected more favorably on domestic production, since some of the most expensive optical components, the ones used in sophisticated military devices, continued to be made in the U.S. In this highly precise field, the U.S. still fared well, supplying about half of DoD's needs in 1986. But the domestic industry was rapidly deteriorating, its employment having dropped from more than 3,000 in 1981 to 1,655 in 1986. The perceived danger was that the Asian countries that had already captured the low-quality mass-produced lens market would soon build on their success to enter the more precise fields of optical production.[82]

Moreover, the potential for a recovery of the U.S. industry, the joint group found, was severely constrained. As the production of lenses, prisms, and other components had migrated abroad, so had the production of optical glass, the raw material for optical components, the manufacture of which required specialized capabilities for melting and blending. There remained only a single domestic maker of optical glass. And while unskilled U.S. labor was plentiful, the number of journeymen and master opticians, persons who had undergone lengthy training, had dropped from 1,015 in 1981 to 602 in 1986. (The number excludes "dispensing opticians" who grind eyeglasses and can become master opticians only after extensive retraining.)[83]

The Joint Logistical Commanders' study group, like the earlier Army Materiel command study, concluded that the domestic production of optical components could not be made to "surge" adequately in the event of emergency mobilization. It recommended that the secretary of defense make a finding that would apply the Federal Acquisition Regulation to precision optics, thereby restricting purchases from foreign producers.

FAR Studies and Ball Bearings

To be sure, this study of precision optics was not the only industrial surge and mobilization study undertaken by the Joint Logistical Commanders. For the very purpose of undertaking such studies, DoD had entered into a Master Memorandum of Understanding with the U.S. Department of Commerce, to have the Commerce Department work with the services to prepare analyses of specific industries. Besides doing the study of precision optics, the Joint Logistics Commanders and the Commerce Department also evaluated the gas turbine, industrial fastener, and ball bearing industries. Separately, working with just the Navy, the Commerce Department analyzed the growth of foreign acquisitions on six types of naval requirements, including missiles, sonobuoys, and communications equipment, in studies that encompassed all stages of manufacturing from raw materials to final products.[84]

Appearing in a Senate hearing on defense industries, Paul Freedenberg, under secretary of commerce for export administration, said that "We at commerce generally agree with the effort to provide more coordination for the issues concerning defense industrial base planning."[85] In his prepared statement, he added that

> The defense industrial base is not a monolith. Each industry is different—facing different problems and requiring different solutions. Although some issues can be addressed at a macro level (e.g. by tax law, antitrust statutes, etc.), most issues need to be examined on an industry-specific basis.[86]

As of early 1988, the Pentagon had agreed to the recommendations of the industrial study of the ball-bearing industry. The project was made the responsibility of Robert Costello, under secretary of defense for acquisition, who stressed that it was not a giveaway but a program of planned industrial recovery. "Some would have us subsidize obsolescence, and we don't want to do that," said Costello in response to a journalist's query about the program. "When they [the members of the industry] lay out a plan to become world-wide competitive, then we'll be there to help."[87]

Optics Rebuffed

While embracing the rescue of ball-bearing firms through the implementation of the FAR, the Pentagon initially rebuffed the proposal for similar treatment of precision optics. On March 2, 1988, William Taft IV, deputy secretary of defense, sent a memorandum directing each of the armed services to reject the Joint Logistical Commander's recommendation. The memo reasoned that the regulation would mainly affect the high-technology precision optics firms, which were still viable, but would do little for the low-cost mass producers, which had disappeared from the U.S. Furthermore, important allies would be hurt by the restriction.[88]

According to coverage in the trade press, the allies were West Germany, long an important manufacturer of precise optics, and the U.K., where one company held a Singaporean subsidiary that supplied the majority of DoD's night-vision devices. Both European countries lobbied against the regulation.[89] Moreover, amid budget deficits, defense agencies would have to be selective about which industries they would attempt to rescue. As quoted in the *Wall Street Journal*, one Air Force officer said "The defense budget is getting smaller, and we're just not going to shield the optics industry from foreign competition."[90]

Taft's memorandum did, however, hold out some hope that DoD would consider assisting firms in plant modernization and encourage apprenticeship training programs.[91] The main lobby group for the optical component industry, the American Precision Optics Manufacturers Association, responded first with continued demands for import restrictions.[92] Simultaneously, in anticipation of the greater salability of alternatives to import protection, the organization changed its strategy.

Decisions were made to shift gears in favor of a research center specializing in manufacturing techniques for precision optics. Published sources vary on the origins of the new strategy. An article in the Rochester press says that the idea can be traced to joint decisions by Duncan Moore, the head of the University of Rochester's Institute of Optics; the head of a Rochester area community college program in optics; and Harvey Pollicove, an Eastman Kodak official who came to be appointed acting director of the new center.[93] Robert Leshne, head of the Precision Optics Manufacturers Association, describes the origins of the proposal differently. He writes that the determination to pursue this option emerged from several meetings held between the association's members and a number of DoD officials and congressional staffers. "The most significant result of those meetings," writes Leshne, "was the conclusion that a future plan of action should be organized to include a joint effort by government, industry, and academia; as for example the system used so successfully over the past several years by the Japanese."[94]

Organizing for the New Center

Contrary to the Japanese model, however, industrial and academic participants, and possibly defense agency officials, but not civilian government representatives, would sit on the new center's board. Moreover, according to Leshne, "the major source of funding for this program will be through the Department of Defense," though he expected significant industrial support by the program's third year.[95]

The manufacturers of optics went out in full force to lobby for the revised rescue strategy. However, the industry had so declined in this country that few optics manufacturers remained. The members of the Optics Manufacturers Association numbered in the forties, at most. Makers of precision optics could not, therefore, match the achievement of the ball-bearing manufacturers, who mobilized calls of support from thirty congressional offices to press the Pentagon for generous terms in restructuring the industry.[96] Yet even the level of lobbying that could be mustered by optics makers had an effect. According to a staff member in the office of Republican Alphonse D'Amato, senator from New York, the senator was approached by representatives of the proposed center, the optics manufacturers, and the Rochester business community. In response, D'Amato joined five other senators in a letter to the secretary of defense in January 1989 expressing displeasure at DoD's rejection of initiatives to build up American optics. The senators urged the secretary to help the industry by setting up an advanced manufacturing technology center sponsored by industry, academia, and government.[97] In his answer to D'Amato, Deputy Under Secretary McCormack said that DoD was again exploring strategies for providing funds to precision optics.[98]

Exerting greater relative leverage, eight members of the House sent another joint letter in May directly to a lower-rung official, the deputy chief of staff for production in the Army Materiel Command. The letter praised the command's announced intention to make the center part of its Optics Thrust Program.[99] In startling concern about a rather specialized corner of military logistics, the representatives wrote that

> Joint technology development activities resulting from the AMC-APOMA-COM [Army Materiel Command, American Precision Optics Manufacturers Association, Center for Optics Manufacturing] partnership will re-establish the American manufacturing base's ability to competitively produce optics for military use.

But the representatives also implied that they and the Army Materiel Command had a common interest in the broader topic of U.S. competitiveness. Their letter concluded that "industry expansion and growth can result only if the industry is competitively positioned with leading edge technology," as if to suggest that the command had a role in maintaining such U.S. technological preparedness.

The Center for Optics Manufacturing

By the summer of 1989, the army indeed committed funds, apparently $13 million, to the Center for Optics Manufacturing for developing machines for the automated manufacture of precise optical components.[100] Soon thereafter, Deputy Under Secretary McCormack instructed the Navy, Air Force, Defense Logistics Agency, and Strategic Defense Initiative Organization to follow the Army's lead in financially supporting the project.[101] Finally, press coverage in 1990 announced a five-year $35-million grant from the Department of Defense in support of the program but cautioned that there was no written commitment and the figure would be reviewed each budget year.[102] Pollicove, the first head of the center, stated that its mission was "to develop advanced optics manufacturing technology, implement technology transfer and establish training and education for the domestic optics industry."[103]

So, by the decade's end, the Center for Optics Manufacturing seemed close to financial viability as a collaborative industrial research center established in part through military planning. In its origins in lobbying by optics firms and its proposed structure as a university-industry research organization, the center also demonstrated disaggregative and collaborative policy at work. But it was a set of studies in the armed services that initially identified the optics industry's competitive plight, and the military agencies' initiatives seem to have inspired the industry's self-organization.

Furthermore, it was military funding on which the program's existence depended. Initiatives to set up the Center for Optics Manufacturing may be characterized, therefore, as being military industrial policy, though a policy on which the Department of Defense itself initially took an ambivalent position.

DARPA AND ADVANCED DISPLAY TECHNOLOGY

Even though the Defense Advanced Research Projects Agency conducts no research itself but rather contracts it out to private firms, universities, and federal laboratories, it is the most glamorous of the military research agencies. Created by law in 1958 in reaction to the launching of Sputnik, DARPA became the agency within DoD that explored radical technological possibilities, ones going beyond the immediate concerns of the armed services and their laboratories. Among defense agencies, it was the one investing in "high risk, high payoff" technologies.

To encourage rapid response to promising technologies, the agency practices a freewheeling kind of management, allowing its program managers considerable leeway in proposing, supporting, and funding projects. In 1989 congressional appropriations and the projects it manages for other defense agencies gave DARPA a budget of more than $1 billion for such research and development. With a billion-dollar exchequer, but a staff of only between 125 and 150, including office support personnel, each program manager had tens of millions to spend on attractive projects.

As DARPA officials make clear, the agency's purpose is research on technologies relevant to military preparedness. But they also readily admit that the military sector is tightly bound with the civilian industrial base and that technologies meant for eventual military use will doubtlessly have industrial worth. Proper technological investments are likely to benefit industrial competitiveness, and industrial competitiveness helps military readiness. Hence, DARPA officials accept that their work does, and should, take industrial effects into account. With such industrial purposes and a massive budget, in the hands of an elite administration, DARPA has often been referred to as America's MITI.[104]

Though the agency is not well known to the public, those who are aware of it attribute to it the origins of some of the most important technologies affecting American industry and daily life. The attributions are indeed startling in that apparently serious observers make them sweepingly, with no provisos. The agency is said to have originated computer time-sharing, high-level computer languages, parallel computing (an important principle in supercomputing), packet switching (essential in computer communications), and computer graphics. As we saw in the passages on the history of photonics, the agency sponsored path-breaking work on lasers. Even a computer specialist said to be relatively unenamored of the agency says that 90 percent of the interesting inventions in computer science in the 1960s, including computer graphics and personal computers, came out of DARPA.[105]

In the 1980s, DARPA's most famous or infamous involvement has been its Strategic Computing Program. It was originally intended to develop computer-mediated responses for battle-field decision making, for pilots engaged in aerial conflicts, and most ambitiously, for strategic decision-making made under the threat of nuclear war. Despite considerable public skepticism about the wisdom of giving computers such a role, even a recent prospectus for the agency speaks of the development of "weapon systems that have human-like reasoning properties."[106] One product of the program, the smart truck that is self-guided through artificial vision, reportedly has been unable to negotiate turns at 2 MPH and has tried to drive up a tree. At the same time, research on the program has yielded a substantial number of artificial intelligence applications now entering commercial use.[107]

DARPA in Photonics

Three of the agency's subdivisions also sponsored research that played a part in the rise of photonics in the 1980s. The Defense Sciences Office had a program officer dedicated to optical and optoelectronic applications in computing. The Directed Energy Office conducted a range of research on laser weapons and countermeasures that could be taken against such weapons. And the Office of Defense Manufacturing took early responsibility for a program on advanced display technology. The programs varied in the extent to which each one sought an industrial effect. (DARPA's internal organizational structure was reorganized in 1991; the office titles used below represent older designations.)

In the Defense Sciences Office, the program in optical computing stressed not optical processing but the more modest aim of using optical fiber to interconnect electronic processing operations. The program was led by John Neff from 1983 to 1988, with a budget of $12 million in his final year. In a 1989 interview,

Neff stated that the office was among the more defense oriented of DARPA divisions. "Other offices were more applications oriented. Industrial questions might play more of a role there."[108] Neff said that in his years at DARPA, during which he handled more than sixty contracts on optical applications in computing, he observed little discussion of the projects' significance for civilian industry, though all welcomed the idea of the technology eventually finding its way to industrial users.[109]

His successor in the position, Andrew C. Yang, similarly said that defense needs came first. "We justify our work by military requirements," explained Yang. "But we all realize that for the military to have strength, we need ample civilian strength."[110] In an administrative environment frequently referred to as "entrepreneurial," decisions on emphasizing or de-emphasizing civilian effects depended greatly on the program manager. "Doing the work of program manager is partly a selling job," said Yang—it is a matter of persuading superiors of the worth of a proposed project. "One selling point," he added, "is that industry finds a need" for the proposed project, but Yang stressed that this was just one selling point among others.

As the program officers readily admitted, one reason for conservatism in predicting industrial applications was that research on optical computing was still far from being feasible in actual computing instruments. Nevertheless, grants announced by DARPA in 1989 facilitated industrial involvement in optical computing. Two of those grants also illustrated the close relationships between DARPA research programs and industrial research having a collaborative form. The two grants were awarded to the Microelectronics and Computer Technology Corporation, a consortium of more than twenty large corporations (such as Lockheed, Eastman Kodak, and General Electric) sharing an interest in computer technology. The grants went to the consortium's Optics in Computing Laboratory to fund the development of holographic computer memory devices and applications for optoelectronics in parallel computing. By mid-1990 the laboratory's researchers were supposed to provide a report identifying the most promising optoelectronic applications for insertion in computer architectures.[111] Though DARPA may well have invested in such work primarily for military purposes, the consortium it contracted with was explicitly set up by its corporate members to reinforce their ability to compete with foreign microelectronics firms.

While the Defense Sciences Office invested in technologies for which industrial uses were yet far off, other offices attended more carefully to the industrial benefits of research. Such projected benefits were stressed quite openly in the part of DARPA's prospectus that described the Directed Energy Office. According to the brochure, "the office's second major thrust is a new initiative in solid state laser and nonlinear optics technology." This research was to help in weapons that disable enemy sensors and in submarine laser communications, "as well as in commercial applications," which were unspecified. Recognizing that such outcomes require concurrent developments in several fields, the office coordinated programs in complementary technical fields "with a commitment to establishing a thorough scientific and engineering base." The passage concluded: "This long-term dedicated technology program protects us against technological surprise from the Soviets as well as the loss of technology edge to the Japanese (MITI) and European (Eureka) manufacturing applications."[112]

The brochure, intended for prospective applicants for DARPA research funds, clearly invited comparisons of the agency to MITI and to a civilian industrial research program of the European Community.

As with other military agencies like SDI, such assertions could not be taken at face value. The agency may have been as concerned about its budget requests and congressional support as about industrial effects. Motivations were convoluted. But in DARPA more than in other agencies, the balance of motivations shifted toward industrial intervention. And within DARPA, it was another office, the Defense Manufacturing Office, that leaned the most in the industrial direction.

The Debate Over Display Technology

Among defense research projects with a relationship to civilian industry, perhaps none received more attention in the late 1980s than DARPA's initiative in high-definition displays. A number of articles, agency reports, and crowded dockets at congressional hearings revealed wide interest in high definition television, so attention naturally turned to DARPA's newly announced effort in the field.[113] It was no secret, indeed it was widely acknowledged, that high-definition TV offered something rare in technology policy proposals: popular appeal.

Mass media reporting on the proposals played on the superior technical characteristics of high-definition television—on the pictures that would be the quality of photographic slides. The reports also frequently mentioned the reestablishment of this country's television manufacturing industry or even its consumer electronics industry. Among those closely involved in the debates, however, the subject occasioned unusually broad thinking about the industrial effects of technological development. Some recognized the relationships of government-regulated telecommunications to futuristic home-entertainment services, especially once homes were wired with fiber optic cable. Others pointed out the importance of a consumer electronics industry in reviving and maintaining domestic semiconductor technology. Still others envisioned future home high-definition entertainment centers that could be integrated with electronic publishing and would, in effect, constitute residential telecomputers.[114]

Robert A. Mosbacher, secretary of commerce, cited such multiple industrial effects in congressional testimony that supported a federal initiative toward advanced TV. He observed that the technology had numerous applications in computer work stations, computer-aided manufacturing, and medical diagnosis equipment.[115] In early 1989, the Bush administration appointed his agency to propose options for the administration's response to advanced television. Options presented in the study included federal investment in joint ventures, specific regulatory exemptions for a high definition TV industry, and government-sponsored standards. But by the fall of 1989 news reports told that the administration appeared to be scuttling Mosbacher's initiative.

According to one unnamed administration official quoted in the *Wall Street Journal*, the initiative "did more than flirt delicately with industrial policy, despite all the (Bush) campaign rhetoric against industrial policy." The newspaper also quotes an administration official's comment that the cabinet-level Economic Policy Council was not likely to approve plans targeted to one industry: "The point has been driven home completely that any proposal must be generic, across-the-board and good for all business—and not directed toward HDTV."[116]

DARPA Meets Display Technology

DARPA, however, had no such scruples about undertaking an initiative in display technology. Early in the same year that the Bush administration appeared to be rejecting such an initiative,

DARPA announced a $30 million research program in high-definition display technology to be administered by the agency's Defense Manufacturing Office. The driving force behind the program, DARPA's Craig Fields, apparently overstepped unwritten rules on permissible explicitness in military sectoral policy. His actions eventually led to his ouster by the Bush administration. But DARPA's programs toward advanced displays and other technologies having critical implications for American industry continued unabated.

To be sure, the programs had primarily military purposes. In military use, high resolution displays would serve in command and control stations for directing battlefield operations, training simulators for pilots and tank commanders, and work stations for analyzing foreign intelligence.[117] By one estimation, the USS *Vincennes*, which shot down an Iranian passenger airliner in 1988, could have avoided doing so if it had been equipped with a color-coded high-definition screen instead of the usual radar screen.[118]

In presenting their program to the public and to Congress, DARPA officials indeed mentioned these military applications but placed far more emphasis on consequences for U.S. industry. A spokeswoman for the agency stated that it was the Pentagon's goal to develop advanced-display technology to revive U.S. consumer electronics and semiconductor industries. "If there was a commercial advanced display industry," she said, "we could fulfill our needs at lower cost." And, she added, "the second thing is that we are very interested in making sure that the domestic semiconductor industry is very robust."[119]

Craig I. Fields, DARPA's deputy director for research and later briefly its director, was even more explicit. Testifying before the House Subcommittee on Telecommunications and Finance, he gave a wide-ranging rationale for military interest in display technology. He stated that the Defense Department "has taken the position that national security requires an assured domestic supply of the highest performance and highest quality technology in our industrial base." He added that DARPA was the department's main instrument for achieving this end.[120]

Of three reasons given by Fields for this position, the first two have already been introduced in this chapter. One was the need for an industrial presence that could be mobilized in wartime. "A second and more compelling reason," he said, was that the Defense industrial base cannot be separate from the much larger volume commercial and consumer industrial base... The economics just don't allow DoD to go its own way, and as the civilian industrial base moves off shore the risk for Defense sharply increases.[121]

The third reason, however, seemed unprecedented in the range of industries it would encompass. "Third and most troublesome," Fields continued,

if our military adversaries turn to the companies in foreign countries for their weapons systems technology, perhaps through theft or surreptitious purchase, and if that technology is more advanced or available earlier than in the US, we will be in serious trouble.[122]

DARPA interest in advanced displays, therefore, had a triple justification, and the third could be used to justify military attention to almost any technologically oriented industry that was more advanced abroad than it was in the United States.

TARGETING DISPLAY TECHNOLOGY

Unlike the cabinet-level discussants of high-definition TV, DARPA saw no problem with, indeed it embraced, sectoral technology policy. With a budget that seemed to allow considerable administrative discretion, the agency was able to put together a $30 million project from internal funds, without having to go to Congress for an appropriation. How did DARPA choose this technological sector?

In early 1988 it was already known in Washington that the Defense Department was interested in the high-resolution graphic capabilities of advanced display technology. In what seemes to have been a precipitating move, Zenith, the last U.S. television maker, and also the maker of a reputable monitor for portable computers, contacted DARPA with a proposal for a research project. Fields and his staff thereupon started inquiries over the desirability of issuing a formal request for proposals.[123]

Fields and his staff visited several labs and met informally with experts. Asked about the procedure used, one staff member of the Defense Manufacturing Office said, "it's a case of talking to different companies, different industries, visiting their labs, and having people come in." Little documentary information was prepared, at least for public release. "We had a one-page fact sheet passed out in our public affairs office," and slide presentations, or viewgraphs, were assembled.[124] DARPA also organized an invitation-only meeting on "photonics," a meeting at which the agency tried to identify the high-priority photonic technologies that would be further investigated.[125] (Results of the meeting were not available to the public.) Viewgraphs were prepared showing information such as statistical trends in the semiconductor industry; Fileds used some of these graphs in his congressional testimony.

In explaining to Congress his agency's selection of advanced display technology, Fields did, however, reveal three criteria for technological choice, criteria that explicitly recognized the multisectoral effects of technology.

Criteria were necessary, he said, because DoD cannot afford to assure supplies of all necessary technologies. As a first criterion, DoD should concentrate on technologies of the greatest military significance. "Secondly, we need to concentrate on technologies that are rapidly improving, so that we suffer a real disadvantage if we get behind even by a few years." And third, "we must focus on technologies that have leverage for a wide variety of end use products, so as to get the greatest benefit from our investment."[126] In the DARPA administration's estimation, then, technologies had qualities—military significance, rate of change, and multiple use—that made some of them more strategic than others.

DARPA announced its solicitation for research on advanced display technology in early 1989. Everyone acknowledged that $30 million was a paltry sum for such purposes. A leading MIT researcher on high-definition TV observed, however, that even though the sum was "peanuts" it stirred the interest of most large computer and electronic companies in the country.[127] The solicitation drew eighty-seven proposals.[128]

The corporate response put DARPA in a position of mediating joint ventures among large private corporations. As Fields stated, "In the HDTV area, I was pleased to see that AT&T and Zenith had joined forces, and we intend to create more such links."[129] Furthermore, Fields asserted, DARPA was unique in federal government in having experience in forming industrial teams, supporting consortia, and having access to a defense industrial technology base unavailable to any individual U.S. company. Indeed, he suggested, DARPA's technological assets would represent a shared resource for the nation's corporations. This technology base "allows us to leverage complementary leading edge technologies ahead of anyone else, gaining precious lead time."[130] By Fields's statements, the agency appeared to have a mediating role often attributed to MITI, that of creating joint corporate relationships that take advantage of complementary technical expertise in government and several corporations.[131]

Cause Célèbre Inside the Beltway

A year after giving his congressional testimony, the year in which he was promoted to agency director, Fields was peremptorily transferred—news accounts used the term "ousted"—from his job. According to the reports, the immediate cause of his removal was a DARPA decision in April 1990 to directly invest $4 million in a small California company, Gazelle Microcircuits Inc., which was designing computer chips out of gallium arsenide. Donald Atwood, deputy secretary of defense, was said to have taken offense because such investment represented an economic intervention that the Bush administration shunned.[132]

Underlying the the event was a long-simmering dispute with the Bush administration, particularly John Sununu, White House chief of staff, and Michael Boskin, chief of the Council of Economic Advisers.[133] According to the *Wall Street Journal*,

> Mr. Fields rankled the White House and some Pentagon officials by leading the government in promoting a federal role in the development of high-definition television, advanced semiconductor manufacturing and a host of other technologies with potentially big commerical payoffs.[134]

News reports agreed that at the heart of the dispute was industrial policy. Fields's ouster became something of a *cause célèbre*, if a small one by Washington standards. The House Armed Services Committee held hearings on his transfer.[135] Eleven members of congress, whom the *Wall Street Journal* described in an editorial as "the entire industrial policy crowd," wrote a letter demanding his reinstatement.[136] But the protests were to no avail. Fields resigned to take up a new position as head of the Microelectronics and Computer Technology Corporation.

His removal from DARPA did not mean, however, that DARPA was turning away from research projects that, to serve military needs, would have to cultivate U.S. industries. The Fiscal Year 1991 budget, a tight one for most agencies, included $75 million for DARPA's advanced display project, $63 million more than the

White House had requested.[137] A congressional provision inserted in the FY 1991 defense budget also gave DARPA $50 million for the development of "pre-competitive" technology.

DARPA set aside $20 million of that for optoelectronics. "The emphasis on optoelectronics appears entirely appropriate," Jeff Bingaman, senator from New Mexico, said in response in 1991. "Every analysis I have seen shows us falling behind in the development of that technology."[138] Early discussions showed that the organizations that would have strong interest in such a consortium included AT&T, Boeing, General Electric, GM/Hughes, IBM, and Lockheed.[139] According to the director of DARPA's Microelectronic Technology Office, one of the new offices set up during reorganization in 1991, this optoelectronic program would lead to the setting up of one or two consortia *in addition* to the three optoelectronics consortia, mentioned earlier in this chapter, that were already being organized as the outcome of the optoelectronic materials initiative. Most of the other projects in photonics conducted in the microelectronics office also consisted of research on dual-use technologies.[140]

Even Donald Atwood, the deputy secretary of defense who is said to have directly brought about Fields's removal, stated in an interview in 1991 that "it is up to DARPA to pick up those technologies that are the wave of the future and make those investments so that we are not militarily dependent on offshore manufacturers." He stressed, as DARPA officials had long emphasized, that DARPA was in the Department of Defense and its first concern in investing in new technology was application to defense products. But "since the overwhelming majority of the technologies we work on do have dual uses, we encourage companies to exploit the commercial side."[141]

In this context, then, Fields's removal represented not a remaking of DARPA but a reassertion of the political formula that became prevalent in the 1980s: industrial intervention occurs in forms that disguise its character. In making clear assertions about DARPA's roles in American industry and in being energetic in pursuing the full ramifications of military industrial programs, Craig Fields overstepped the boundaries, exceeding the degree of forthrightness within which military industrial programs could be carried out or discussed. According one report, Victor Reis, Fields's successor at DARPA, was expected to maintain DARPA's traditional stance. "If he does his job right," said one Senate staffer of the new DARPA director, "there's enough latitude to support spin-off technology that he doesn't need to be a lightning rod."[142]

Indeed, DARPA has continued to have at its disposal an unmatched budget for affecting the disposition of industrial technology. To independently select uses for the funds, the agency needs to be secure from political winds. DARPA employees prize the relative political calm in which they operate. This sentiment was revealed at the end of a *New York Times* feature article about DARPA. The reporter asked a former director of the agency why DARPA's functions could not be carried out by a civilian agency. He replied, "I don't think the mission is clear enough on the civilian side." The reporter then finished the thought as follows: "That would make a civilian agency prone to politics and bureaucratization,"[143] evils from which DARPA's projects are relatively sheltered.

MILITARY PLANNING FOR U.S. INDUSTRY?

In programs addressing optoelectronic materials, optics manufacturing, and advanced displays, defense agencies conducted sectoral policy. Such policy, by definition, leads us to the problem of knowledge in sectoral choice. How indeed did the military agencies make decisions supporting some technologies rather than others? Was the military sector better at planning and strategy than was the civilian sector?

According to the principled opponents of industrial policy, the fundamental difficulty in such policy is the problem of criteria. How does one decide which sectors deserve special governmental attention? The question partly mis-specifies the problem, since judgments can reflect a decision maker's ongoing interactions and professional experience, as well as hard criteria. Yet it properly raises the question of the intellectual grounds for policy choice.

Military practitioners of industrial policy seem to find no great difficulty in describing their criteria. Officials concerned about such matters readily offer simple rules of thumb by which sectors can be chosen.

In announcing DARPA's decision to pursue a program on advanced displays, Craig Fields mentioned three rules for technological selection. The technology should (1) have great military significance, (2) be undergoing such rapid change that the U.S. could quickly fall behind without an investment, and (3) have the potential for many end uses, so that the investment pays off in numerous applications.[144] Speaking at hearings of the Committee on Armed Services, even William Graham, the Reagan administration's director of science and technology policy, readily listed four similar criteria by which "to prioritize and share efforts in advanced and exploratory research." He included the availability of research teams and the effectiveness of competing technologies among his criteria.[145]

Paul Freedenberg, under secretary of commerce, also specified the criteria by which his agency worked with the military services to identify industries requiring special military attention. The department's studies judged "the short and medium-term capabilities of specific defense related industries to respond to increased production requirements under peacetime, surge and mobilization scenarios." In studying such capacities of a specific industry, such as precision optics, the agency gave particular consideration, Freedenberg stated, to industrial competitiveness, foreign sourcing and dependency, production constraints, trade patterns and market trends, and labor requirements.[146]

The criteria that DARPA and other military agencies used for industrial and technological selection were indeed only simple rules of thumb. They begged numerous questions about how we classify technologies and take account of their interdependencies. They were criteria of the most rudimentary kind. Yet such simple criteria for making technological choices were at least considered in military-industrial policy-making, while they appeared very rarely in civilian (disaggregative and collaborative) policies toward photonics.

Within DARPA, officials and program managers also had ready access to specialists in the technical fields under consideration. The agency could arrange confidential conferences on selected technologies, staff tours of research facilities, and visits with prominent researchers. DARPA staff members could make policy judgments not only with simple criteria but also through a variety of consultations with technical specialists.

Technology Lists and Forecasts

While the military industrial programs used rudimentary criteria to make technological decisions, other branches of the military establishment had access to more substantial means of evaluating military technologies. The greater depth of thinking about technologies meant for military purposes is reflected in the phenomenon of the technologies list and, more significantly, in extensive technological forecasting exercises.

Decision making with lists of critical technologies is reasonable in principle, since a decision to give greater priority to some rather than other technologies implies some overview of the alternatives. Lists are, however, a problematic technique. They beg the essential questions. How do we classify technologies? How do we differentiate among the several levels at which we deal with technology? (For example, information technology, low-energy photonics, photonic information processing, optoelectronic circuitry, gallium-arsenide, and gallium arsenide fabrication techniques each represent progressively narrower levels at which we understand a technology.) And there are many other questions. Without braving the conceptual challenges, defense agencies prepared several lists in the 1980s, each differing somewhat from the others, each also having significant overlaps with the others.[147]

In one of the most encompassing of the publicly available lists, the DoD Critical Technologies Plan, technologies were selected for inclusion by their potential for enhancing weapons systems and providing new military capabilities.[148] The plan occasionally mentions nonmilitary technological achievement in the U.S. and abroad, but only in passing. Domestic industrial capabilities in technology were not a criterion for selection.

In its 1988 Summer Study, the Defense Science Board took exception to the restrictive criteria by which the lists are prepared. The lists emphasize weapons and their performance "and overlook general industrial knowhow," the board wrote. Sensor technology is a strong example of the "coupling" between industrial and military technologies. Gallium-arsenide sensors in particular could be "critical to either military sensor suites or industrial process control." The domestic capacity to apply technology in industry, the board suggested, is as important as the more strictly military criteria for inclusion on the list.[149] However, such broader, industrially oriented listings of military technologies did not appear in the 1980s.

In technological planning, military agencies have also produced something much more substantial than the technology list. They have conducted elaborate forecasting exercises projecting their future technological need. They include Navy 21, Strategic Technology for the Army, and Air Force Forecast II.[150] Results are classified, but some second-hand reporting on the Air Force forecasting project gives us a sense of how it was conducted.

Completed in 1986, Forecast II identified thirty-nine new technologies and thirty-one "systems concepts" that could transform the Air Force's capabilities through the early twenty-first century.[151] More than 175 Air Force and Pentagon staff members participated in discussions, along with other persons from government, industry, and universities. Three panels evaluated ideas and suggestions. The Technology Panel included subpanels organized by engineering discipline. The Mission Panel comprised subpanels devoted to strategic defense and offense, low-intensity warfare, battle management, and theater warfare. And the Analysis Panel traded off ideas and concepts in subpanels that examined relative costs and relative threats.[152]

We should not be surprised that Forecast II identified photonics, by that name, as one of the most strategic of the technologies. Photonic technology was seen as essential in battle-management systems, communications and computing systems that would operate if damaged, and large arrays of sensors, as well as specific devices for robotics vision, "smart weapons," and cockpit displays.[153]

At a news briefing on Forecast II's results, an Air Force spokesman stressed that the service would be "aggressively" incorporating the project's results into science and technology planning. The Air Force was already redirecting 25 percent of its science and technology account toward programs urged by the forecasting exercise.[154] One action the Air Force took was to establish a Photonics Center in 1987 at Griffiss Air Force Base in Rome, N.Y.

Two years later, photonics researchers at Griffiss still credited the forecasting project with having coordinated Air Force work on optics. "We always have had optical activity going on," a researcher said, "but Forecast II really pulled it together." They also gave credit to the exercise for giving visibility to photonics, protecting it at a time of cutbacks in Air Force research spending.[155]

Though Forecast II invited the participation of specialists from the industrial world, it did not seem to have examined U.S. industrial capability in the new field. From the reporting, one gathers that industrial concerns played a role in Forecast II but did not elicit serious study. And while the results of the forecast were apparently used to rationalize Air Force research investments, no such studies seem to have been used to lend coherence to military projects directed at industrial technology. Therefore, the most substantial technological planning exercises in the armed services appear to have operated quite separately from military programs directed at industry. The military industrial programs themselves apparently did not conduct such forecasting studies.

Coordination and Planning

If the results of Forecast II were not integrated with military industrial programs, the oversight reflects less a prejudice against industrial programs (though that, too, exists) than a more general incoherence among military technological programs. The Air Force may have used forecasting to set its own technological priorities, but chances are that it did not coordinate them with those of the Navy, Army, DARPA, SDI, the Office of the Secretary of Defense, and the intelligence agencies. The problem is well enough known. By 1987 no fewer than seventeen studies had exposed problems of planning, management, and coordination among DoD science and technology programs.[156] In 1989, the Office of Technology Assessment again observed:

> DoD lacks a strong and focused coordinating capacity for its science and technology programs. Although DoD has over 200 tri-Service and inter-agency coordinating groups, in general they have not been effective in providing high level coordination across the DoD-wide technology base programs.[157]

The blue-ribbon committees that studied DoD industrial base policy restricted their scope to DoD's industrial problems but made similar observations. They, too, recommended more management and coordination. By the end of the decade, some of the recommendations were being put into effect.

In 1988, DoD was reorganized to consolidate acquisitions and industrial responsibilities under a new office, that of the under secretary of defense for acquisition. A Defense Manufacturing Board, modeled on the Defense Science Board, would serve the under secretary with industrial advice. Speaking of his mission, Robert B. Costello, the new under secretary, said, "Our findings indicate it is vital that we develop industrial plans to support military strategic plans."[158]

Since 1989, a new Office of the Deputy Under Secretary of Defense (Production Base and International Technology) has been responsible for "acquiring, organizing, maintaining, and disseminating data requested for defense industrial base plans and programs."[159] Two rather ambitious databases are available for DoD's industrial efforts. One called "Socrates," still classified as of 1989, is said to enable DoD to "identify industrial technologies critical to maintaining a robust, internationally competitive industrial base." It is said to track capability, technology by technology, country by country, in number of years that it is ahead of or behind the U.S. The other system, the Defense Industrial Base Network, reportedly monitors U.S. industries, identifies essential military suppliers, and tracks their ability to rapidly increase the production of critical items. And a Strategic Planning Task Force has been established to develop a DoD Industrial Strategic Plan.[160]

DoD seems to be on the verge of systematization in its dealings with the industrial base. Or is it? Congressional staff members are, as ever, skeptical, feeling that DoD has not genuinely taken industrial issues to heart, but has only engaged in administrative reshuffling. "The defense industrial base is still an orphan there," said Ed McGaffigan of Senator Bingaman's office.[161] It would remain to be seen in the next decade whether reorganization and computerization had given DoD the ability to plan coherently for industry.

In all, through the 1980s, the defense agencies did bring to bear in industrial decisions a broader and richer set of considerations than did policy-making that took a disaggregative or collaborative form. Yet the defense agencies never incorporated into industrial decisions the breadth of strategic thinking that went into technological planning for more traditional military purposes.

CONCLUSION

Military Industrial Policy

The preceding sections have portrayed defense agency industrial programs directed at photonic technologies. The sections reviewed SDI's investment in optical computing, DoD's research in optoelectronic circuitry, the evolution of DoD support for a research center in optics manufacturing, and DARPA's program in advanced displays.

Do these constitute industrial policy? Our earlier definition serves as only a partial guide. As defined, industrial policy operates through sector-specific knowledge and judgment about the needs and opportunities of specific sectors. But nearly all of military acquisitions require awareness of the productive potential of specific industries and affects those industries through the scale of military buying power. Military R & D also requires an awareness of the wider world of technological development and, unless the research is classified, has an effect on that technological world. It seems excessively all-encompassing to refer to military acquisitions and military research in themselves as industrial policy. If they were so understood, too much in the military world would qualify under this definition.

For discussions of military policy, the definition should be more stringent. Military industrial policy, therefore, should qualify as such only when it undertakes to shape a domestic industrial or technological sector to some purpose, such as productive capacity, ability to mobilize in war, or ability to trade successfully in an internationalized economy. And the policy should be more than an intention; it should be reasonably capable of affecting the conditions of production in an industry.[162]

By this definition, SDI's effort in optical computing cannot be characterized as industrial policy. During the period considered, research on optical computing was still far from being realized in practice. Outside of a short-lived optical computing consortium, no actions were taken that would affect the conditions of production in industry. Promises of spinoffs were a public relations claim or, at best, a distant ambition.

The three other programs do, however, qualify as industrial policy. To build domestic capacity in optoelectronic materials, DoD invested in pilot plants and joint ventures for the production of gallium arsenide chips. Though the agency's primary purpose was to lower costs for military purposes, it chose to do so partly through a research program that would foster domestic production for commercial purposes as well. In actively seeking to create an industry, DoD's optoelectronics initiative became industrial policy.

In the wake of the studies of the joint logistical commanders, DoD sought to influence the manufacture of precision optics largely to facilitate wartime mobilization. By 1990, DoD decided to do so by supporting a research center on manufacturing methods in precision optics. Since the program sought to strengthen the ability of the manufacturers to survive in world markets, this initiative, too, was industrial policy.

In allocating funds to build industrial capacity in advanced display technology, DARPA sought primarily to serve the diverse military uses for such displays in aerospace, battle management, and surveillance. But the agency saw such military ends as completely intertwined with the development of commercial markets. Seeking to simultaneously affect U.S. technological standing in both realms, DARPA became an arbiter of sectoral technological investment. Surely that was industrial policy. Therefore, even by our more stringent definition, military agencies instituted distinct policies aimed at American industrial capacity in photonics.

Policy-Making Knowledge

By definition, such policy-making raises questions of sector-specific knowledge. What were the conceptual means by which the technological sectors were chosen? Through the 1980s, the military industrial programs influencing photonic technology emerged from disparate political and bureaucratic sources and only the most rudimentary thinking about civilian industry.

Despite the recognized importance of optoelectronic materials to military strategy, and the shortfall in U.S. commercial capability in the field, one proposal for a research center on the topic was implemented only after the fortuitous convergence of pork-barrel appropriation, DARPA persistence, representations by industrialists meeting in an engineering society, and the intervention of a leading senator. The evolution of the Center for Optics Manufacturing began with a study by the Joint Logistical Commanders of the services, who first recommended a protectionist policy. Only after DoD's initial rejection of protection for optics, and upon further pressure from an industrial lobby, did the center come into existence with military support.

DARPA was different, since the agency's administrators themselves had the authority and were allowed the discretion to directly sponsor research projects on display technology and other fields of photonics. And though the agency could articulate its criteria for the choice and did engage in reasonably extensive consultations before the decision, the choice largely reflected the agency's tradition of speculative technological bets.

Such programs were coordinated neither with each other nor with the other industrial programs established during the decade. Though rule-of-thumb criteria were considered and some studies were conducted, the industrial programs directed toward photonics arose as a clutter of discrete choices.

The armed services did conduct elaborate technological forecasting exercises to set their internal science and technology funding priorities. But such exercises apparently did not address the capability of U.S. industry to fulfill the services' forecasted demands. In making sector-specific technological choices, the military agencies took into consideration broader sets of criteria and evidence than did civilian technology policy, but through the 1980s the military agencies had not yet incorporated their industrial programs into a broader strategy for the technological development of American industry.

Privatization

Privatization generally refers to the transfer of activities from government to the private sector, but if we only slightly stretch the meaning, it also encompasses the relegation of public activities to an isolated, enclosed sphere, where ideologically uncomfortable activities can be pursued in a more agreeable guise. Did initiatives toward the industrial and technology base constitute a privatized industrial policy?

Our answer cannot be one-dimensional. In organizing industrial programs, the military officials' own motivations were convoluted. Perhaps military technology initiatives created an occasion for industrial policy. Then again, promises of industrial spinoffs helped protect cherished military programs from budget cutters. One senses in the military discussions of industrial policy less an intent to dissemble than genuine contradictions and ambivalence.

And an answer probably resided less in the motives of the military officials than in the purposes of those who funded the DoD industrial programs. Led by Jeff Bingaman of New Mexico, the Senate Subcommittee on Defense Industry and Technology operated only for the sake of investigating economic interrelationships among civilian and military sectors. Though the senators themselves did not say so explicitly, they heard frequent testimony that military industrial strategy might do for U.S. industry what civilian policies could not.

For example, Jacques Gansler testified on the need for a comprehensive defense industrial strategy and for the establishment of a separate defense budget entirely for the development of the next generation of manufacturing systems. Chairman Sam Nunn asked him, "Would you do this through the Defense Department?" Gansler replied, "I would do this in the Defense Department primarily because we do not have any other way of doing it. Right now, the United States does not have a MITI."[163]

Congress was well aware of the ideological issues. As a member of Jeff Bingaman's staff pointed out, "There is a lot of opposition from free marketeers in having government target technology." He added, "DoD is by default our technology policy because of the ideology that no other department can pick winners."[164] When operating within the military agencies, industrial policy satisfied both the corporate interests worried about competitiveness, and that long-standing congressional and industrial establishment that supports military appropriations. Their alliance made a solid coalition.

Most of all, industrial programs ensconced in the military establishment were surely sheltered from broader public accountability. Military agencies sought or claimed effects on industrial competitiveness but were never held to task for outcomes. And, inevitably, military priorities and perspectives colored all such policy. The restricted vision in the military establishment partly explains why industrial policies flourished there. In obscure military settings, industrial policies were protected from the broader public interest in the technological directions of the economy.

EIGHT

Conclusions and Implications

Privatization Unbound

Among the responses to the industrial challenges of the 1980s, perhaps none were more characteristic of the decade than the calls for partnership. *Partnership* became the byword for relationships between government and industry, industry and university, and competing industrial firms themselves, not to mention labor and management.

Surely this was a curious ethic, since it was frequently proclaimed by those who, in other contexts, would espouse the more traditional values of competition, the contest of interests, and checks and balances. The leaders of industry, government, and university seemed to have become unexpectedly converted to sanguine notions of human motivation. The economic future, it seemed, rested on cooperation and not unbridled rivalry. But the outcome of these calls to partnership was not a gentle age of mutual trust. The outcome was the relegation of decision making to private, unaccountable realms.

At the beginning of the next decade, the spirit of privatization in American policy-making was by no means exhausted. It seemed even more vigorous, having recently taken the unadorned form of "industry-led policy." Promoted by lobbying groups and industrial consortia, industry-led policy would bring about closer government and business collaboration in choosing technologies for governmental assistance, but—in the words of *Business Week*—"with industry calling the shots."[1]

Industry-led policy truly represents privatized public policy in the sense in which we have understood it here. Rather than transferring government assets to private firms for allocation under the constraints of a market, it shifts policy-making itself to shadowy forums that are unaccountable to the broader public concern about the technological directions of the economy.

This concluding chapter reviews how this country's public responses to the rise of photonics have taken such privatized forms. It recapitulates the argument that, operating apart from formally granted authority, unable to plan, privatized policy cannot by its very nature respond to those integral features of technology that warrant a policy response. And it examines some implications of this argument for policy research and policy making.

Coherent Technology

This study has argued that in the 1980s photonics became a technology with an industrial importance on a scale attained by only a few others in this century, among them petrochemical technology, internal combustion, electrical power, microelectronics, and biotechnology. Photonics built on its origins in optics, lasers, fiber optics, and reconnaissance systems to become something approaching an integrated body of knowledge and engineering practice.

It entered widespread practical use as systems of fiber-optic communications were laid throughout the country in the 1980s. Optical-communications networks became the first nationwide systems of photonic infrastructure. Other light-wave systems were becoming feasible. Home and office optical information storage, industrial sensing, high-definition displays, and optoelectronic circuitry all came into use. And their integration into still larger systems was on the horizon.

Photonics now impinges on the affairs of thousands of manufacturers of photonic components and the makers of products that utilize them. It reorients manufacturers toward the sensors and inspection systems through which automation occurs. It redirects numerous firms toward the new imaging, communications, sensing, and computing functions that have become possible. It restructures industrial sectors ranging from copper wire to chemical photography, from aerospace avionics to consumer electronics, and from computers to medical equipment. Photonics has emerged as a challenge to and resource for the productive activities of numerous firms in varied industries.

Photonics has such broad effects by dint of its integral characteristics. It manifests this integrality in three respects: as an emerging body of knowledge and skill, in systems of interconnected devices, and through sets of technological interrelationships with industry.

These integral characteristics of the technology allow us to comprehend and, in some measure, anticipate the development of photonics. Integral properties allow overviews, generalizations, and comprehensive visions. Such properties make the implications of technological change potentially knowable to policymakers. To the extent that a nation or region uses such knowledge to cultivate technological resources, resident firms gain advantages that would be unavailable to them were they reacting to the technology only through discrete actions within a market. The integrality of technological development, therefore, warrants a broadly informed, comprehensive policy response.

Yet, in the United States, despite the domestic origins of much of photonic technology, and a domestic R & D investment that, at least through the mid-1980s, equalled or exceeded that in Japan, industrial performance proved to be mediocre or worse. American industries could not effectively take advantage of the new technology. Plausibly, then, the country's industrial retrogression reflected an incapacity for coherent policy response, an inability to plan.

In utmost contrast to this picture of technology, the conventional discourse in the United States has assumed that technologies, such as photonics, arise as an accumulation of disparate innovations. Technological change occurs in the aggregate of individual discoveries and inventions. Decisions on investment in research are, therefore, best made by those managers, entrepreneurs, and investors closest to unique technical problems and business conditions. An industrial policy that would hope to outguess the professionals would be foolhardy. American industrial effectiveness will, in this view, revive only through the market, since only within it can individuals properly respond to the complexities of technology.

The contest between the two visions hinged on conceptions of technology. This study has suggested the idea of the integrality of technology on grounds of plausibility and not proof. The notion of technological discreteness, for all its conventionality, was equally unproven. The conventional discourse on the U.S. response to technological and industrial change relied, therefore, on a set of implicit and unexamined assumptions about technology.

Those corporate managers, military strategists, and academic scientists who deal daily with photonics did, nevertheless, notice and respond to the nation's lag in the technology. Yet the conventional opposition to planning in the economy and the absence of conceptual tools for grasping the coherence of technological change required a rejection of an explicit policy response. The outcome was a sublimated policy, one that could operate without being explicitly acknowledged. The outcome, in short, was privatization.

Incoherent Policy

Privatization, as we have seen, did not mean the surrender of public policy in favor of market solutions but rather the relegation of policy-making to a private realm. This response was so pervasive that one is at a loss to find research programs that have not been privatized. Indeed, the descriptions in the previous three chapters present a more or less full inventory of government-inspired programs of the 1980s oriented toward U.S. industrial standing in photonics. With partial exceptions within the military establishment,[2] the programs operated in the absence of coherent thinking about the selection of technologies, effects on industries, and the public ends being sought.[3] Though a number of collaborative programs received government funds, none had to account for the industrial purposes of their decisions. A catalog of such programs, therefore, does more than suggest plausibility; it adds up to strong evidence that privatization occurred.

As we saw in the work of Theodore Lowi, Don Price, and H. L. Nieburg, privatization was not a new phenomenon in American politics. In much of agricultural and urban policy, privatized forms of policy-making represented the very essence of the interest-group politics of the postwar era. In the cold war confrontation over advanced armaments and space capabilities, U.S. administrative agencies and research contractors came to be intertwined in a power structure that shielded itself from public accountability. During the 1980s, the nation's industrial problems combined with the illegitimacy of industrial policy to send the U.S. economy inexorably drifting toward a privatized response. The privatization of industrial policy did not, however, simply represent the success of parochial interest groups in dominating policies that impinged on their individual interests. Privatized industrial policy sought to revive U.S. industrial performance in world markets. It hoped to do so by privately appropriating public policy-making in order to influence the very conditions of industrial production.

In a variety of programs directed at photonics, we observed this result when we asked how allocation decisions were made and agendas were set. We observed privatized policy-making in the form of disaggregation, collaboration, and sheltering.

The disaggregative form of privatization responded to economic problems by dispensing government assets to private interests. One of its manifestations, the technological pork barrel, allocated research facilities through back-room budgetary negotiations and specialized lobbying. Another manifestation, business participation in the setting of one agency's research, became the model by which other federal agencies, too, were to adapt their research priorities to corporate preferences. And the National Science Foundation's project selection revealed a third manifestation. Research funding choices were driven down to committees of anonymous academic and industrial peers who made technological judgments for public ends but engaged in no deliberative process and had no discernible source of information to guide them in doing so. Disaggregative policy parcelled out government assets unit by unit, in the absence of any plan or strategy on how these units related to each other or to the broader technological currents in the economy.

Through the collaborative form of privatization, policymakers hoped to respond to U.S. industrial lethargy by establishing industrial consortia and industry-university collaborative research centers that would serve as centers of technological strength. In the industry-university centers, academic members were to determine their research direction through interactions with industrial affiliates. Research priorities were to be set through collaborative committees composed of academic and industrial members—and with no governmental agenda-setting role.

The collaborative research centers intermingled the purposes of academic and industrial members such that decision making emerged from a subtle balance of power. In some centers, industrial affiliates managed to extract cut-rate contract research from the university, while in others the affiliates agreed to let faculty members themselves set the research agenda in accordance with academic predilection. In neither case did the collaborative research centers built any coherent source of information or basis for sound judgment by which they could respond with vision to the public purposes for which they had been established.

In the third form of privatization, industrial technology programs were pursued in the shelter of the military agencies. The programs varied greatly. Some were mere public relations claims to industrial benefit from military programs. Others undertook to shape a domestic industrial sector to increase its productive capacity, foster its ability to mobilize in war, or help it adapt to an internationalized economy. The most significant military bureau in these matters, the Defense Advanced Research Projects Agency, sought to build civilian technological strength primarily to serve military ends, but it also asserted that such ends were hardly distinguishable from those of assisting the domestic technological response to foreign industrial rivals.

Therefore, there did indeed emerge in the military establishment a set of policies that sought effects on industrial competitiveness. In its emphases and choices, military sectoral policy always had to give precedence to military priorities. And while industrial ends could be claimed, they were never made an explicit mission, so that the policy could operate without ever having to demonstrate effectiveness. Industrial policy flourished under the aegis of the military agencies as long as it did not raise the ire of the opponents of economic intervention and kept its programs sheltered from broader public concern with and scrutiny of the technological transformation of the economy.

The privatization of industrial policy, then, proceeds in three forms. But these forms ought not suggest a typology. Disaggregation, collaboration, and sheltering should be seen not as formal types but as headings representing the preponderant theme among a cluster of overlapping characteristics. While the headings suggest the primary character of a privatized program, the programs also frequently take on intermediate and combined forms. Pork-barrel appropriations lead to the establishment of collaborative centers and military research programs. Military bureaucracies establish industry-university research centers and private consortia. The three forms of privatization are mutually reinforcing for avoiding the glare of explicitness and accountability in policy making.

These privatized industrial programs came into being in the 1980s, a time of Republican dominance in the executive branch of the U.S. federal government. But privatization was by no means just a phenomenon of Republican politics. In the growth of earmarked science appropriations, Democrats and Republicans were equally eager participants, ever willing to approve science facilities in one district for favors in kind to their own. The National Institute of Standards and Technology was proclaimed a model of proper business-government relations in the 1980s, though it had long set its agenda through business participation. The National Science Foundation's anonymous decision-making process was transformed from traditional peer review (with ends only in the technical quality of research) to merit review (with social and economic purposes) in the 1970s.

The National Cooperative Research Act of 1984, which eliminated most antitrust barriers to industrial research collaboration, was a fully bipartisan achievement, receiving unanimous approval on both sides of the aisle. And again at the federal level, the Senate Subcommittee on Defense Industry and Technology, which was controlled by the Democrats, gave even more support to military initiatives toward industry than did the Republican administration itself. In New York State, the legislative reorientation of the state science and technology agency to promote collaboration occurred through bipartisan agreement. The privatization of industrial policy was, therefore, not merely a reflection of party politics. This response to industrial dislocations was structurally ingrained in American capitalism.

In all its forms, privatized policy could not respond to the integral characteristics of photonics. The budget negotiations, diffuse business links with government laboratories, merit review, industry-university research committees, industrial consortia, and industrial programs operating under military auspices all made sector-specific technological choices for the sake of international industrial rivalry. But none could give a reasoned explanation of how the choices fit into a broader vision for the economy.

The United States performed poorly in taking industrial advantage of photonic technology because its industrial technology policies were incapable of responding to the integral properties of technological change. Excessive privatization led to the ineffective development of a technology that was the common resource of industrial firms. To the extent that U.S. public policy responded similarly to common resources other than photonics, the U.S. industrial deterioration of the 1980s was in good measure the outcome of incoherent policy, policy that operated without vision or plan.

Implications for Research and Theory

In proposing the idea of the integrality of massive technological change, this study has challenged the conventional opposition to industrial policy. The idea is offered only in rudimentary form. It invites policy research and theoretical elaboration. At stake are significant implications for policy.

Indeed, if the conception holds in contexts other than technology, then resources with integral properties are broadly important in capitalist industrial production. Then surely national and regional industrial performance depend to a degree on policy-making that can monitor, understand, and respond to the changing features of productive resources in the economy. The contest of firms in an unfettered market then turns out to be a less capable form of capitalism. The ability of firms to produce most efficiently might now depend on the state's ability to plan.

The idea of integrality, therefore, provides us with a preliminary argument for industrial policy. To strengthen the argument, we need a theory of common resources that elucidates the integrality of technology.

To that end, we need research on the comparative structures of technological revolutions—biotechnology, superconductivity, advanced ceramics, and microelectronics, in addition to photonics. Indeed, indications are widespread in the reporting on these technologies that pork barrel, anonymous merit review, consortia, industry-university collaboration, and military-industrial programs (but programs of the National Institutes of Health in the case of biotechnology) have consituted the primary U.S. policy responses for the sake of industrial competitiveness.[4]

We also need research that assesses and compares the varieties of common resources, not just technology, and the ways in which policy responds to them. For example, in the face of economic problems in the 1980s, localities throughout the United States responded to changing infrastructure needs through a variety of "partnerships" among land developers, city governments, and public authorities.[5] Similarly, federal, state, and local policy responded to evolving needs for technologically advanced education and job training by encouraging partnerships among local government, colleges, and representatives of private firms.[6] We might find from studies of this response that, amid widespread industrial change, U.S. public policy reacted to problems of infrastructure and technical education much as they did to technological research, through privatization.

Implications for Policy

When massive technological change, such as the development of photonics, transforms the economy, corporations, military agencies, and universities respond with plans and strategies. These organizations do not in themselves take account of the interrelated effects of the technology on various sectors of the economy. A policy that arises from the sum of their actions would fail to take account of the integrality of the resource. Nations and regions excessively dependent on such disparate technological decision making would experience a slow and debilitating industrial decline in comparison to other nations and regions that were better organized to respond to the evolving disposition of their productive resources.

Yet in this country federal and state governments, which do have the authority for a more comprehensive response, act through haphazard appropriations, interest-group pressure, an ambiguous politics of collaboration, and projects sheltered from public accountability. They are perpetuating a failed industrial policy.

For successful industrial policy, federal and state governments must develop the capacity for policy-making that respects the values of coherence and accountability. When directed at technological change, such policy should operate through an ability to monitor emerging bodies of technical skill and knowledge, emerging sets of technical systems, and the technology's developing interrelationships with industries. The policy should reflect a capability for prudence, vision, and judgment about disposition of economic resources that cut across the interests of individual firms. But it would be facile counsel to elaborate here on the practical means to such planning. The ability to make such policy judgments should emerge through professional experience and educated discipline. These are not the stuff of concluding remarks.

The search for planning and coherent policy for U.S. industrial revival is a paradoxical endeavor. As we have seen, privatization in American policy-making is not an occasional failing; it is structurally ingrained. In this study, as in Theodore Lowi's conception of interest-group liberalism, planning stands as the alternative to muddled, privatized technology policy,[7] but the prescription to plan challenges the central premise of the diagnosis. The political economy of privatization would seem to augur the political incapacity to plan.

We may draw some relief from the paradox if we briefly return to the Japanese counterexample. By the end of the 1980s, Japan had, as we have seen, built up the most successful optoelectronics industry in the world.

For distant observers, assertions about Japanese industrial policy are indeed perilous. Yet, to find value in the Japanese example, we need not assume that one agency such as MITI was dominant or that Japanese economic agencies orchestrated their actions. Our more important lesson from Japan may be that a set of traditions and institutions has enabled economic agencies to attend strategically to the multifarious relationships between technology and industries. Whether working in concert or as adversaries, the Japanese agencies seem to have been able to respond broadly to the integral characteristics of technological change.

If we were to try to replicate a Japanese-style economic ministry, we might find that it would flounder, since it would be operating in isolation from a broader institutional framework. We might better ask how the U.S. might build on domestic institutional foundations a capacity for coherent response to technological changes that revolutionize industry.

In the United States, capability for responsible planning could be built through reform of the structures of privatization. The reform should be centered in an agency and deliberative body that engages in policy research and controls the budgets that directly affect research choices. In carrying out this work, the agency should build on the kind of privatization that has most characterized U.S. technology policy in the 1980s—industry-university cooperation and industrial consortia. The agency should enlist collaborative groups to represent their knowledge and perspective in the drafting of national or regional technological directions, while the agency should exert on them the discipline of policy analysis, shared visions, financial controls, and public accountability.

For present purposes, we need only conceive of this agency as an organizational capability in federal or state government. For now, we need not define its formal status as a line agency, foundation, executive office, civilian version of DARPA, or independent body (like the Federal Reserve). We should, however, see it as being quite distinct from the National Science Foundation or the Congressional Office of Technology Assessment.

In contrast to the NSF, which must respond to an academic constituency, the proposed agency should have the discretion and budget to set its sights on the nation's technological needs and to shape technological capacities in research institutions, not just to respond to those institutions' pressures. And in contrast to the

Office of Technology Assessment, which advises Congress, the proposed agency should do more than conduct studies. It should have the authority to make discretionary research funding decisions, as well as to conduct policy study and analysis. At the same time, the agency should formulate broad visions and undertake strategic action through ongoing interactions with industrial consortia and industry-university research centers, on whose boards the agency would be represented.[8]

Indeed it is an organizational structure that combines macroscopic vision and diverse participation that would prevent the well-known flaw of concerted planning. Planning, if poorly organized, could operate through rigid criteria and procedures, impose unitary conceptions on policy, and restrict diversity and experimentation. In contrast, dispersed research centers and the simultaneous operation of numerous visions maintain the responsiveness of U.S. technology policy, even if, in the presence of an internationalized economy, the sum total of such actions falls short of effective policy.

Industrial technology policy must, to an extent, come to terms with the inescapable tension between diversity and independent volition on the one hand and comprehensive vision and accountability on the other.[9] In the United States, the tension has been resolved almost entirely in favor of diversity and fragmentation at the expense of accountability and broad vision. And this country is paying a price for that choice in the decline of its technology-intensive industries.

American industrial technology policy should operate, then, through an administrative structure that creates a flexible accommodation between cohesive policy direction and independent initiative. To plan technological development effectively, an agency should combine both broad planning and situated insight, both accountability and diversity, in industrial technology policy. Collaborative groupings, established with governmental funds or through governmental antitrust exemptions, would be reincorporated into an accountable policy making process. With the agency's representatives appointed to their boards to maintain accountability, the collaborative groupings could form the institutional framework for a workable industrial policy in the United States.

As this study has argued, such capacity for planning is warranted even by strictly economic and industrial criteria. But technologies like photonics also have implications—good ones and bad ones—for education, privacy, public health, and the natural environment. Through the capacity to plan coherently, government policy could guide the development of technologies that not only revolutionize industry but affect the natural environment and public health. An agency that plans public technological investments, combined with a network of public representatives on research boards, could come to exert a measure of control over the technologies that periodically transform our lives. But because of that very possibility, proposals like the present one will be strenuously resisted.

Notes

ONE

1. Hazel Henderson, "The Age of Light: Beyond the Information Age," *The Futurist*, Jul.–Aug. 1986, p. 56; "And Now, the Age of Light," *Time*, Oct. 6, 1986, p. 56.

2. For full citations on data presented in this chapter, please turn to chapter 2.

3. National Research Council, *Photonics: Maintaining Competitiveness in the Information Era* (Washington, D.C.: National Academy Press, 1988), p. 8.

TWO

1. Among book-length technical descriptions for novices, the one that is the most readable, though restricted in its coverage, is Michael I. Sobel, *Light* (Chicago: University of Chicago Press, 1987). M. Y. Han's *The Secret Life of Quanta* (Blue Ridge Summit, Pa.: Tab Books, 1990), introduces readers to the formidable topic of quantum electronics. Several books are available on lasers, including Breck Hitz, *Understanding Laser Technology* (Tulsa, Okla.: Pennwell, 1985) and Clayton Hallmark, *Lasers: The Light Fantastic* (Blue Ridge Summit, Pa.: Tab Books, 1979). A book apparently meant to introduce photonics to teenagers, but written so simply as to be almost incomprehensible, is Valerie Burkig, *Photonics: The New Science of Light* (Hillside, N.J.: Enslow, 1986). A well-explained and well-illustrated but more technical introduction to optics is William P. Ewald, W. Arthur Young, and Richard H. Roberts, *Practical Optics* (Pittsford, N.Y.: Image Makers of Pittsford, 1983). Adequate introductions can also be found in the standard encyclopedias.

2. Pierre Aigrain, "La Photonique, technique de demain," 1973, cited in A. A. Sawchuck, "Society Planning Committee Recommends Name Change," *Optics and Photonics News*, Feb. 1990, p. 16. The place of publication of Aigrain's piece is not given in Sawchuck's citation.

3. The word photonics was used in the 1970s in a different sense to refer to high-speed photographic instrumentation. A discussion of this earlier sense of the word in comparison to its new sense appears in Hallock F. Swift, "Photonics: A Magic Word?" *Photomethods*, May 1982, p. 74. Instances of the early appearance of the word are Richard A. Shaffer, "Scientists Start Testing Ways to Expand Photons' Potential," *Wall Street Journal*, May 15, 1981; Hughes Aircraft Company's corporate organ, *Vectors*, offered an article about photonic logic in its Nov. 4, 1981, issue.

4. The definition used in this study is amalgamated from several sources. *The Photonics Dictionary*, 34th ed. (Pittsfield, Mass.: Laurin Publishing, 1988) defines the word to mean "the technology of generating and harnessing light and other forms of radiant energy whose quantum unit is the photon." It is no coincidence that the definition is very close to that of *optics* in the fourth edition of the *McGraw Hill Dictionary of Science and Technology*: "The study of the phenomena associated with the generation, transmission, and detection of electromagnetic radiation in the spectral range extending from...about 1 nanometer to about 1 millimeter." The latter dictionary has no entry for *photonics*.

Discussions of *photonics* sometimes define the word by analogy with electronics as "an art in which photons take over some of the duties of the electron" (Stewart E. Miller, "Lightwaves and Telecommunications," *American Scientist*, Jan.–Feb. 1984, p. 67). A similar definition is used in the National Research Council's report, *Photonics: Maintaining Competitiveness in the Information Era* (Washington, D.C.: National Academy Press, 1988), p. 1. But the analogy with electronics does not do full justice to photonics, whose applications in directed energy (through the laser) and in capturing images do not have a counterpart in electronics.

A discussion of these matters appears in Brian J. Thompson, "Education in optics—challenges at hand," in *Proceedings of the Society of Photo-Optical Instrumentation Engineers*, vol. 978 (Bellingham, Wash.: 1989). To Thompson, "optical science and engineering" covers all that photonics does and more. For present purposes, however, *photonics* better serves the purpose, not because of technical specificity, but because it better connotes the revolutionary newness of the technology.

Those interested in photonics should be wary of pitfalls in the common usage of related terms. *Optoelectronics* is the most important, since nearly all technical applications combine optical and electronic properties. *Optoelectronics* is often used as a substitute word for the entire range of phenomena described in this book. Note that Japanese and British documents use the term broadly to include fiber optics, while American usage often restricts the word to the components at either end of the optical cable that convert optical signals to and from electronics.

The word *electro-optics* is not generally synonymous with *optoelectronics*. It has a restricted technical definition as the branch of physics that studies the effects of an electrical field on the optical properties of matter. However, trade publications, especially those dealing with a military audience, commonly use *electro-optics* loosely to refer to a range of devices that capture and display images or otherwise use light or images in sophisticated technical devices. *Optronics* is more or less similar in meaning. The term *lightwave technology* is usually but not exclusively associated with fiber-optic communications and related work on optical computing and optical circuitry.

5. See Anthony J. DeMaria, "Photonics vs. Electronics Technologies: 1988 Ives Medal Address," *Optics News*, Apr. 1989, pp. 22–37.

6. A review of laser technology and its applications may be found in Jesse H. Ausubel and H. Dale Langford, eds., *Lasers, Invention to Application* (Washington, D.C.: National Academy Press, 1987).

7. See Sanford Lakoff and Herbert F. York, *A Shield in Space? Technology, Politics, and the Strategic Defense Initiative* (Berkeley and Los Angeles: University of California Press, 1989), chap. 3.

8. On the relatively small size of the laser industry, see "The Missing Boom in Lasers," *New York Times*, Nov. 13, 1986, p. D1.

9. See National Research Council, *Photonics*, chap. 9.

10. A survey of such developments appears in Jun Shibata and Takao Kajiwara, "Optics and Electronics are Living Together," *IEEE Spectrum*, Feb. 1989, pp. 34–38.

11. For more information on these broad applications of photonics, see National Research Council, *Photonics*, chap. 3 and 4.

12. A summary article covering the technical issues and disputes is Trudy E. Bell, "Optical Computing: A Field in Flux," *IEEE Spectrum*, Aug. 1986, pp. 34–57.

13. Popular articles on optical computing and its economic potentials include Jeff Hecht, "Optical Computers," *High Technology*, Feb. 1987, pp. 44–49; Tom Alexander, "Computing with Light at Lightning Speeds," *Fortune*, July 23, 1984, pp. 82–88; and "Pushing Computers Closer to the Ultimate Speed limit," *Business Week*, July 28, 1986, pp. 48–49.

14. For more on sensors, see National Research Council, *Photonics*, chap. 5. An article on smart skins is "Materials That Think for Themselves," *Business Week*, Dec. 5, 1988, pp. 166–67.

15. See Leonard E. Ravitch, "Electronic Imaging: Promises and Threats," *Laser Focus*, Jan. 1986, pp. 128–36.

16. National Research Council, *Photonics*, considers displays to be photonic devices. See chapter 4.

17. For example, during the years 1983–88, the investment house E. F. Hutton (later Shearson Lehman Hutton), New York, produced a series of reports and collections of conference presentations entitled *Electronic Imaging Forum*. The series was devoted to highlighting the potentials of investment in electronic imaging.

18. Current thinking about high-definition television recognizes the relationship with fiber-optic communications. See, for example, U.S. Department of Commerce National Telecommunications and Information Administration, *Advanced Television, Related Technologies, and the National Interest* (Washington, D.C.: 1989), p. 31. An article on imaging written for general audiences is "The Graphics Revolution," *Business Week*, Nov. 28, 1988, pp. 142–56.

19. L. R. Baker, "The Precision Optical Industry," *Optics News*, Feb. 1989, pp. 13–17.

20. "History of the Optical Society of America, 1916–1966," *Journal of the Optical Society of America* 56, no.3 (Mar. 1966): pp. 2–3.

21. On the history of the University of Rochester program, see Hilda G. Kingslake, *The Institute of Optics, 1929–1987*, Rochester, N.Y.: University of Rochester College of Engineering and Applied Science, 1987.

22. "History of the Optical Society of America."

23. Advisory Council on Science and Technology, *Optoelectronics: Building on Our Investment* (London: Her Majesty's Stationery Office, 1988), p. 65.

24. "SPIE: Thirty Years of Development and Applications of Optical and E-O Science and Engineering," *Laser Focus/Electro-Optics*, Aug. 1984, p. 10.

25. This passage provides only a thumbnail history of the laser. A thorough history of the laser through 1970 became available as the present manuscript was being completed. It is Joan Lisa Bromberg's *The Laser in America, 1950–1970* (Cambridge, Ma.: MIT Press), 1991.

26. Opening Statement by Senator Schmitt, in U.S. Congress, Senate Committee on Commerce, Science, and Transportation, *Laser Technology, Development, and Applications*, Hearings before the Subcommittee on Science, Technology and Space, Serial No. 96-106 (Washington, D.C.: U.S. Government Printing Office, 1980), p. 100. (Henceforth this cited as *Laser Hearings*.) Some of the early history of quantum physics is recounted in Sobel, *Light*.

27. Larry Miller, "The Laser on the Battlefield," in David Fishlock, ed., *A Guide to the Laser* (New York: American Elsevier, 1967), pp. 68-84.

28. Reminiscences of these years appear in Charles H. Townes, "Harnessing Light," *Science '84*, Nov. 1984, pp. 153-55.

29. Joan Lisa Bromberg, "Research Efforts That Led to Laser Development," *Laser Focus*, Oct. 1984, p. 60.

30. "Now the Father of the Laser Can Get Back to Inventing," *Business Week*, Feb. 17, 1986, p. 98; and David Kales, "Laser Pioneer Completes 30-Year Odyssey for His Patents," *Laser Focus*, May 1988, p. 96.

31. Bromberg, "Research Efforts That Led to Laser Development," p. 60.

32. Bromberg, "Research Efforts That Led to Laser Development," p. 58. Detailed work on the military history of laser technology may be found in two articles in the journal *Historical Studies in Physical and Biological Sciences*, 18, no. 1 (1987). They are Robert Seidel, "From Glow to Flow: A History of Military Laser Research and Development," pp. 104-147; and Paul Forman, "Behind Quantum Electronics: National Security as Basis for Physical Research in the United States, 1940-1960," pp. 150-229.

33. Statement of George F. Smith, in *Laser Hearings*, p. 100.

34. Smith statement, in *Laser Hearings*, pp. 99, 104-5.

35. Statement of George Gamata, in *Laser Hearings*, p. 90.

36. Gamata statement, in *Laser Hearings*, p. 90.

37. Garry W. Gibbs, "U. S. Inertial Fusion Program," *Laser Focus*, Oct. 1984, pp. 200-4.

38. T. P. Long, "Laser Processing—From Development to Application,," *Living Case Histories of Industrial Innovation* (New York: Industrial Research Institute, 1981), pp. 38-40.

39. Bromberg, *Laser in America*, pp. 113-143, 225. The data on numbers of firms are from page 114.

40. Bromberg, *Laser in America*, pp. 120ff, discusses these events in the early days of Spectra Physics and other laser firms.

41. In constant 1973 dollars, the total output of the laser industry for sales to non-communist countries grew from $244 million in 1973 to $480 million in 1979. Over that time, sales to government consistently accounted for more than 60 percent of all sales. Of sales to government, most were sales to the military. Sales to government's nonmilitary branches (14 percent of total sales in 1979) went mainly for research on fusion. See "Review and Outlook, 1979," *Laser Focus*, Jan. 1979, p. 37.

42. *Laser Focus*, Jan. 1980, p. 32. At least some divisions of the military establishment have felt that all the laser investment was worth it. The director of the Laser Division of the DoD Night Vision and Electro-Optics Laboratory made a strong claim for the military value of lasers in Senate testimony in 1980, though the claim must be seen in the context of his pointed remark to the senators that the application of laser systems is "constrained by limited funding and not by technology." "Low energy laser development," he said, "has been the single most successful DoD investment in the last decade." See Gamata statement, in *Laser Hearings*, p. 99.

43. There was wide agreement after Desert Storm that the war had been good for investment in sophisticated weaponry. See, for example, John D. Morrocco, "Gulf War Boosts Prospects for High-Technology Weapons," *Aviation Week and Space Technology*, Mar. 18, 1991, pp. 45–47.

44. *Fiberoptics explained* (Newport, R. I.: Kessler Marketing Intelligence, 1985). A more detailed explanation of fiber optics and a review of its early history can be found in Sami Faltas, "The Invention of Fibre-Optic Communications," *History of Technology* 5 (1988): pp. 31–49.

45. C. David Chaffe, *The Rewiring of America: The Fiber Optics Revolution* (Boston: Academic Press, 1988), pp. 4–5.

46. Faltas, "The Invention of Fibre-Optic Communications," p. 42.

47. J. M. Senior and T. E. Ray, "Optical-Fibre Communications: The Formation of Technology Strategies in the UK and USA," *International Journal of Technology Management* 5, no. 1 (1990): pp. 71–88. The approach to Corning Glass is discussed on p. 74.

48. L. C. Gunderson, "Optical Waveguide—Matrix Management Met Complex Technical and Commercial Problems" *Living Case Histories of Industrial Innovation* (New York: Industrial Research Institute, 1981), pp. 13–17.

49. See Gary K. Klauminzer, "Twenty Years of Commercial Lasers—A Capsule History," *Laser Focus*, Dec. 1984, pp. 54–79; Tingye Li, "Lightwave Telecommunications," *Physics Today*, May 1985, pp. 24–31; Statements of George Gamata and Joseph A. Giordamaine, in *Laser Hearings*, pp. 54, 94; Chaffee, *Rewiring of America*, p. 9. Narratives of AT&T's role in fiber-optic development are in Jeremy Bernstein, *Three Degrees Above Zero* (New York: Charles Scribner's Sons, 1984).

50. Calvin Sims, "Fiber Optics: The Boom Slows," *New York Times*, Nov. 13, 1988, p. D8.

51. Bonnie Feuerstein, "FOG-M, A Giant among Military Fiberoptic Programs," *Laser Focus*, Aug. 1987, p. 98.

52. Giardamine statement, *Laser Hearings*, pp. 64–65.

53. Bernard J. Bulkin, "Imaging Science—a Discipline Emerges," *Laser Focus*, July 1985, p. 14.

54. Richard D. Schwartz and David S. Nelson, *The Growth and Development of the Electronic Imaging Industry: Opportunities and Challenges* (New York: Shearson Lehman Hutton, 1987), p. 1.

55. Public Broadcasting System spoke openly of the agency in a program that first aired on Oct. 13, 1987. See WGBH Educational Foundation, "Spy machines," [Transcript of television program "Nova" #1412] (Boston, Mass.: WGBH Transcripts, 1986).

56. James Bamford, "America's Supersecret Eyes in Space," *New York Times Magazine*, Jan. 13, 1985, pp. 38–39, 50–54.

57. Bamford, "America's Supersecret Eyes in Space."

58. Bamford, "America's Supersecret Eyes in Space."

59. A history of space espionage and its technologies can be found in William E. Burrows, *Deep Black: Space Espionage and National Security* (New York: Random House, 1986). Reference to the origins of the charge-coupled device is on p. 249.

60. Bruce R. Mackie, "Military Imaging: Image is Everything," *Laser Focus*, Aug. 1988, p. 148.

61. Mackie, "Modern Military," p. 148.

62. Forman, "Behind quantum electronics." Aside from Forman's fine article, a brief reflection on the context of the emergence of quantum electronics is in Charles H. Townes, "Ideas and Stumbling Blocks in Quantum Electronics," in Frederick Su, ed., *Technology of Our Times: People and Innovations in Optics and Optoelectronics* (Bellingham, Wash.: SPIE Press, 1990), pp. 68–75.

63. *The Photonics Buyers' Guide* (Pittsfield, Mass.: Laurin Publishing, 1990), cited in Robert E. Fischer, "Healthy Growth in Optics," *OE Reports*, May 1990, p. 2. Another indicator of the scale of U.S. interest is *The Corporate Technology Directory*, U.S. ed., vol. 1. (Wellesley Hills, Mass.: Corporate Technology Information Service, 1987, p. xvii), a commercially produced guide that seems to go to lengths to seek out what the editors define as high-technology products. As of the directory's 1987 edition, it includes photonics & optics as one of its major categories, along with seventeen other major categories, such as advanced materials, biotechnology, and test & measurement. In the 1987 directory, photonic products consisted of acousto-optical equipment, cameras, displays, optoelectronic devices, fiber optics, magneto-optic equipment, optics, and related equipment. The 12,370 independent entities (almost all of them businesses) listed that year produced about sixty thousand products, of which 5.5 percent (thirty-three hundred products) had Photonics & Optics as their primary designation.

64. These firms' entry into fiber optics, except for that of IBM, are described in Carol Suby, "Fiber Optics: Not Just a Niche Market Any More," *Electronic Business*, Feb. 1, 1986, pp. 28–30. The information on IBM is from a news item in *Photonics Spectra*, Feb. 1989, p. 44.

65. K. Ritrievi et al., *Minnesota Mining and Manufacturing Co., Company Report* (New York: Paine Webber, 1988).

66. Claudia H. Deutsch, "Kodak Pays the Price for Change," *New York Times*, Mar. 6, 1988, sec. 3, p. 1; "Air Force Awards Xerox $10 Million Contract," *Democrat and Chronicle* (Rochester, N.Y.), Jan. 8, 1988.

67. Schwartz and Nelson, "Growth and Development," p. 20.

68. Optoelectronic Industry and Technology Development Association, *Annual Report 1987* (Tokyo, 1988), p. 2. In the information available in English, no documentation is offered on how the data were collected. The data seem to be in inflated, not constant, yen. The Japanese figures may well be understating the commercial impact of photonics. The figures exclude many image capture and display devices, sensors, certain lasers, and lenses and other optical components (some of which might not be considered optoelectronics, but are part of photonics). The figures also omit a host of products meant for military use, though these are more likely to be produced in countries other than Japan.

69. Notwithstanding statistical sources of diverse origin and dubious accuracy, this note estimates the scale at which goods incorporating photonic technology are produced. The least controversial conclusion one can draw from the data is that commercial markets are already big.

Notes to Chapter 2   247

Though market data consultants make it their business to project demand for the future, even data on the past is hard to come by. For the year 1986, three separate sources happen to be available and tend to corroborate each other.

Consider that, for 1986, Japan's Optoelectonic Industry and Technology Development Association estimated the demand in five industrialized nations for a selection of optoelectronic products at close to $5 billion (*Annual Report 1987*). To obtain a figure that encompasses the full range of photonic products, one should add to this amount the markets for imaging products and systems, sensors, a range of electro-optical devices (such as night-vision and surveillance equipment) for military use, lasers and optical products not included in the Japanese study, and commercial sales in additional countries. If all these were added, the worldwide figure for 1986 might be double the demand estimated by the Japanese. Such a figure would match *Time* magazine's unattributed estimate that advanced optical technology would yield a worldwide market size of $10 billion in 1986 ("And Now, the Age of Light," *Time*, Oct. 6, 1986, p. 56).

Trying to anticipate the importance of optoelectronics for proposed technology policies for the United Kingdom, the Advisory Council on Science and Technology also tried its hand at estimates in its 1987 publication *Optoelectronics*. Relying mainly on market research firms and other secondary sources, the Council attempted to represent the importance of the technology by guessing the optoelectronic content in the products of several economic sectors. It projected that the technology would account for large segments of business in telecommunications equipment, information systems, consumer electronics, military acquisitions, automotive electronics, aerospace avionics, medical equipment, materials processing, industrial process control, safety and security, and energy.

The council, however, gave up the attempt for the consumer electronics, safety and security, and aerospace avionics sectors. It did estimate that the value of optoelectronic content was $1 billion in the medical equipment sector, $3.2 billion in the Western world's military sector, and approximately $4 billion in the telecommunications-equipment sector. The council's estimates were no model of statistical consistency, making some estimates for 1986 and others for 1987. Nevertheless, the scale of the estimates tends to corroborate a figure of $10 billion for total worldwide photonics sales in 1986.

Since growth rates worldwide have not been as fast as those in Japan, it seems reasonable to estimate that by 1989 the value of commercial sales worldwide (including Japan) was in the range of $15–20 billion.

70. Again the sources are disparate. The U. S. Department of Commerce, which acknowledged biotechnology but not photonics, estimated in

the *U.S. Industrial Outlook 1989* (Washington, D.C., 1989), p. 19–3, that the value of "biotechnology-driven" products in the United States stood at $1.2 billion. The figure in Japan was also near $1 billion in the same year, according to Neil Gross, "Japanese Biotech's Overnight Evolution," *Business Week*, Mar. 12, 1990, pp. 59, 72. If we attribute a generous $800 million to the rest of the world's biotechnology production, the world total stood at only $3 billion in 1989.

71. Optoelectronic Industry and Technology Development Association, *Annual Report 1987*, pp. 5, 6. Figures are rounded. Data on Japan's share of world markets for optoelectronic products would better make the point that Japan has been becoming dominant in the field, but no reliable data seem to be available in that form.

72. Optoelectronic Industry and Technology Development Association, *Annual Report 1987*, pp. 6 and 7.

73. United States International Trade Commission, *US Global Competitiveness: Optical Fibers, Technology and Equipment*, USITC Publication 2054 (Washington, D.C., Jan. 1988), p. 7–1.

74. National Research Council, *Photonics*, pp. 65–66.

75. U.S. International Trade Commission, *US Global*, chap. 8.

76. Len DeBenedictis, "Foreign Competition Poses Threat," *Laser Focus World*, Jan. 1989, p. 19.

77. James W. Phelps, "Competitiveness in the U.S. Optics Industry," *Optics News*, Apr. 1989, pp. 7–11, data are on p. 8.

78. National Research Council, *Photonics*, p. 8.

79. The foreign shares of U.S. patents in lightwave communications (48 percent) was larger, but not remarkably larger, than the foreign shares of all technologies (46 percent) in 1986. The data came from a set of U.S. Patent and Trademark Office studies, *Technology Assessment and Forecast Report*, issued periodically, undated, and available on microfiche.

80. Science Applications International Corporation, *JTech Panel Report on Opto and Microelectronics*, U.S. Dept. of Commerce Contract No. TA–83–SAC–02254 (Springfield, Va.: National Technical Information Service, 1985), p. xv.

81. W. T. Tsang, "A Bird's Eye View and Assessment of Japanese Optoelectronics: Semiconductor Lasers, Light-Emitting Diodes, and Integrated Optics," in Science Applications International Corporation, *JTech Panel*, pp. 5–1 to 5–22.

82. "The High-Tech Race—Who's Ahead," *Fortune*, Oct. 13, 1986, pp. 29–37.

83. Tassey compiled the data by surveying U.S. and foreign industry officials and observers of the industry. He asked about the numbers of scientists and engineers that perform optoelectronics research; his survey, in effect, measured the *performers* not the funding *sources* of research. He projected estimates to 1986. See Gregory Tassey, *Technology and Economic Assessment of Optoelectronics* (Gaithersburg, Md.: National Bureau of Standards, 1985).

84. Data on the perecentage of R & D perfomed by academic institutions may be found in National Science Board, *Science & Engineering Indicators—1987* (Washington, D.C.: U.S. Government Printing Office, 1987), pp. 78, 238.

85. The figure was compiled from U.S. Department of Defense program element codes for DoD expenditures for research, development, and advanced development in programs judged to be closely related to electro-optics in military agencies excluding the Defense Advanced Research Projects Agency and the Strategic Defense Initiative Organization. Categories related to electro-optics consisted of the following: airborne radar imaging, computing (electro-optic), displays, fiber optic communications, fiber optic guidance, fiber optics (other), optical interconnects, optical processing, sensors, and electro-optic systems.

86. In the same year, 27 percent of Department of Defense R & D was performed in government intramural labs (National Science Foundation, *Federal Funds for Research and Development: Fiscal Years 1986, 1987, and 1988*, vol. 36 [Washington, D.C., n.d.], p. 13). If the same proportion is applied to the $230 million, then $53 million can be added to optoelectronics research taking place under government auspisces, with more to add for SDI and DARPA.

87. Compiled from Department of Defense program element codes for research, development, and advanced development.

88. The funding consisted of $10.7 million handled by the Defense Research Advanced Projects Agency, $3.8 million by the Air Force Office of Scientific Research, $3.3 million by the Strategic Defense Initiative, and $680 thousand by the Army Research Office. See Bell, "Optical Computing," p. 353.

89. According to published sources, a U.S. Central Intelligence Agency document purports to estimate American, Japanese, and Soviet R & D expenditures related to optics. The classified report is said to state that the United States spent $1 billion on optics-related research

(apparently in 1985), compared to $3 billion spent by Japan and $4–10 billion spent in the USSR. (The information is found in "And Now the Age of Light," *Time*, Oct. 6, 1986, p. 56, and Bell, "Optical Computing: A Field of Flux," p. 53., though the latter, in what seems to be a mistake, refers to the amounts as expended on optical computing research.) For Japan, a figure of $3 billion in 1985 would form about 8 percent of that year's total national R & D expenditure (National Science Board, *Science and Engineering Indicators—1987*, Washington, D.C.: U.S. Government Printing Office, 1989, p. 234, according to 1985 exchange rates). Not only does such an estimate seem excessive, it far exceeds figures cited by Tassey's report. A Soviet expenditure of the range suggested verges on the fantastic. It may be that the CIA had good intelligence only on the U.S. If so, my point on large domestic R & D expenditure in photonics is reinforced.

90. Compiled from Trudy E. Bell, "Japan Reaches beyond Silicon," *IEEE Spectrum*, Oct. 1985, pp. 46–52; Jonathan Joseph, "How the Japanese Became a Power in Optoelectronics," *Electronics*, Mar. 17, 1986, pp. 50–51; and "Industry Association Promotes Japanese Optoelectronics Development," *Laser Focus*, Feb. 1984, pp. 104–5.

91. Seiji Hagiwara, "Creating Tomorrow's Information Technology," *Journal of Japanese Trade and Industry*, no. 3, (1987): pp. 15–18.

92. Clyde Prestowitz, "Japanese vs. Western Economics," *Technology Review*, May–June 1988, pp. 27–36.

93. U.S. Congress, Office of Technology Assessment, *Holding the Edge: Maintaining the Defense Technology Base*, OTA–ISC–420 (Washington, D.C.: U.S. Government Printing Office, 1989), p. 164.

94. Office of Technology Assessment, *Holding the Edge*, p. 164.

95. Chalmers Johnson, *MITI and the Japanese Miracle: The Growth of Industrial Policy, 1925–1975* (Stanford, Calif.: Stanford University Press, 1982).

96. Chalmers Johnson, "The Institutional Foundations of Japanese Industrial Policy," in Claude E. Barfield and William E. Schambra, eds., *The Politics of Industrial Policy* (Washington, D.C.: American Enterprise Institute for Policy Research, 1986), pp. 187–205.

THREE

1. Thomas S. Kuhn, *The Structure of Scientific Revolutions*, 2nd ed., (Chicago: University of Chicago Press, 1970).

2. Kuhn, *Structure of Scientific Revolutions*, p. 10.

3. Kuhn, *Structure of Scientific Revolutions*, p. 11.

4. Kuhn, *Structure of Scientific Revolutions*, p. 17.

5. For this observation, which appears in *Structure of Scientific Revolutions*, (44 n. 1), Kuhn cites Michael Polanyi, *Personal Knowledge* (Chicago, 1958).

6. "The members of all scientific communities," Kuhn wrote, "including the schools of the 'pre-paradigm' period, share the sorts of elements which I have collectively labelled 'a paradigm.'" Kuhn, *Structure of Scientific Revolutions*, p. 179.

7. An example of the dilemma encountered when all science is seen as pardigmatic is the following: "What changes with the transition [of a paradigm] to maturity is not the presence of a paradigm but rather its nature. Only after the change is normal puzzle-solving research possible" (Kuhn, *Structure of Scientific Revolutions*, p. 179). But now an established scientific tradition is defined in terms of its being normal and established.

8. A compendium of such scholarship is Giovanni Dosi et al., eds., *Technical Change and Economic Theory* (London: Pinter, 1988).

9. Giovanni Dosi, "Sources, Procedures, and Microeconomic Effects of Innovation," *Journal of Economic Literature*, 26 (Sept. 1988): pp. 1120–71, esp. p. 1127.

10. Giovanni Dosi, "Sources, Procedures, and Microeconomic Effects," esp. p. 1127.

11. Dosi's proposed definition is barely helpful even if one manages to disentangle the punctuation. He writes:

> A 'technological paradigm' defines contextually the needs that are meant to be fulfilled, the scientific principles utilised for the task, the material technology to be used. In other words, a technological paradigm can be defined as a 'pattern' for solution of selected techno-economic problems based on highly selected principles derived from the natural sciences. A technological paradigm is both a set of *exemplars*—basic artefacts which are to be developed and improved (a car—of the type we know—an integrated circuit, a lathe, etc., with their particular techno-economic characteristics) and a *set of heuristics*—'Where do we go from here?', 'Where should we search?', 'On what knowledge should we draw?', etc, (consider, for example, general rules of the kind: 'strive for an increasing miniaturization of the circuit,'...).

The passage, with stress as in the original, is from Giovanni Dosi, "Nature of the innovative process," in Dosi, et al., eds., *Technical Change and Economic Theory*, pp. 221–39. The passage appears on p. 224.

12. Dosi, "Nature of the innovative process," p. 226.

13. C. Freeman, J. Clark, and L. L. G. Soete extensively explain their dependence on the Schumpeterian tradition in *Unemployment and Technical Innovation* (London: Pinter, 1982). The classic reading on gales of creative destruction is Joseph A. Schumpeter, *Capitalism, Socialism, and Democracy* (1942; reprint, New York: Harper and Row, 1975), chap. 7.

14. Christopher Freeman, *Technology Policy and Economic Performance: Lessons from Japan* (London: Pinter, 1987), p. 64.

15. "Industry briefs," *Photonics Spectra*, Aug. 1989, p. 62.

16. These research centers are listed and discussed in chapter 6.

17. A schematic on optical communications devices, applications, and integration appears in Optoelectronic Industry and Technology Development Association, *OITDA Activity Report for Fiscal Year Ended March 31, 1989*, vol. 2 (Tokyo: 1989), p. 14. Also see Advisory Council on Science and Technology, *Optoelectronics: Building on Our Investment* (London: Her Majesty's Stationery Office, 1988), p. 48.

18. Edith W. Martin, "Photonics: A Lever to Change the World," *Photonics Spectra*, Jan. 1991, pp. 84–85.

19. On the sociotechnical systems created by electrification, see Thomas P. Hughes, *Networks of Power: Electrification in Western Society, 1880–1930* (Baltimore: Johns Hopkins University Press, 1983).

20. Not all commentators subscribe to the nicety of this distinction. For example, Perez and Soete use the term *system* to distinguish the conceptually interconnected nature of products and inventions. They write (with reference to work by Freeman) that "products build upon one another and are interconnected in technology systems. Each product cycle develops within a broader family which in turn evolves within an even broader system...This means that each 'new' product benefits from the knowledge and experience developed for its predecessors and its producer profits from the already generated externalities." The authors then use "system" in the sense that "paradigm" is used in the present text. See Carlota Perez and Luc Soete, "Catching up in technology: entry barriers and windows of opportunity," in Dosi, et al., eds., *Technical Change and Economic Theory*, pp. 458–479. Quoted passage is on p. 475.

21. For example, components for optical communications are produced in telecommunications firms, notably AT&T, and optoelectronic materials are made by military contractors such as TRW. It is not always possible to distinguish the industry making the technically innovative components usable in many other industries from the industry making complex final products.

22. Just one example of downstream technical effects can be taken from the credit card industry. The company American Express supplanted receipt slips with laser-printed sheets containing images of the original receipts. As told in a *New York Times* article (John Markoff, "American Express Goes High Tech," July 31, 1988, sec. 3, p.1),

> At the heart of the change are new technologies that are revolutionizing the way American Express handles mountains of paper and billions of pieces of data that are its stock in trade. The key technology involved in the billing—image processing—converts all that paper and information into digital images stored on computer disks. The images can be recalled instantly in readable form, or effortlessly transmitted thousands of miles away.

23. A French literature has emerged to analyze such industrial interrelationships around a technology. It refers to the interrelationship as a *filière*, which literally means 'thread.' Brief descriptions of the concept in English can be found in Margaret E. Sharp, ed., *Europe and the New Technologies* (Ithaca, N.Y.: Cornell University Press, 1986), pp. 135–41; and Stephen S. Cohen and John Zysman, *Manufacturing Matters: The Myth of the Post-Industrial Society* (New York: Basic Books, 1987), pp. 101–2.

24. The employment data reflect information on Monroe County, the jurisdiction that includes Rochester. The source is *County Business Patterns* (Washington, D.C.: U.S. Department of Commerce, Bureau of the Census, 1986).

25. A coffee-table book celebrating Kodak's contributions to photography admits the potential threat to chemical photography. See Douglas Collins, *The Story of Kodak* (New York: Harry N. Abrams Publishers, 1990), pp. 374–85. The new imaging technologies and their effects on the photographic industry are described in Peter Behr, "Does Kodak Have Any Home-run Hitters," *Los Angeles Times*, Sept. 11, 1986; and Phil Ebersole, "Imaging is Changing the Kodak Picture," *Democrat and Chronicle* (Rochester, N.Y.), Aug. 19, 1984, p. 1F. The story is much more complicated. Kodak has hedged its bets partly by increasing its own investments in optoelectronics research but more by making substantial investments in pharmaceuticals and

biotechnology, subjects seen as more directly related to traditional corporate strengths in chemistry. Kodak might therefore manage the technological transition, but it might set its new technical directions in areas other than Rochester (new pharmaceutical investments have gone to the Philadelphia area), so that the future of its traditional home is less assured.

26. They are not under the control of individual firms, unless they are monopolies. But monopolies contradict the assumption that we are speaking here about competitive markets.

27. A new technological paradigm causes dislocations that are not captured by conventional industrial codes, such as those guiding the statistical collections of the U.S. Bureau of the Census. The Census Bureau industrial codes do not bring out, indeed they obscure, the complex web of interconnections that a technological paradigm weaves among industries. The bureau collects industrial data according to standard industrial classifications (SICs). How does it respond to photonics? It does so by classifying older optical industries that fall under readily usable headings, such as Optical Glass, Ophthalmic Goods, and Optical Test and Inspection Equipment. The most recent revision of the classification still omits—or does not explicitly include—imaging equipment and systems but does have codes that include optical scanning, optical storage, and optical disks within broad categories that also contain nonoptical products. For example, optical recorders and scanners are under SIC 3577, Computer Peripheral Equipment Not Elsewhere Classified. The conventional information does not provide means by which industries can be examined for their technological interrelationships. See Executive Office of the President, Office of Management and Budget, *Standard Industrial Classification Manual* (Washington, D.C., 1987).

28. To be sure, this logic is expressed here in excessively general terms. More specific policy guidance should emerge from research on technological integrality. Such research should examine the forms that integrality takes, whether as bodies of knowledge and skills, technical systems, technological interrelationships among industries, or other forms. From such research, we should learn how to identify specific conditions where macroscopic technological decisions are likely to be superior to the decisions of individual market actors.

29. U.S. Congress, Office of Technology Assessment, *Holding the Edge: Maintaining the Defense Technology Base*, OTA–ISC–420 (Washington, D.C.: U.S. Government Printing Office, 1989), chap. 9.

30. The word *planning* is used here in a very general sense to refer to intelligent choice in collective decision making (choice based on

research, professional experience, informed judgment, deliberation, and debate), as opposed to decision making that arises from the disparate self-seeking choices of market actors.

31. Lenny Siegel and John Markoff, "High Technology and the Emerging Dual Economy," in David Bellin and Gary Chapman, eds., *Computers in Battle: Will they Work?* (Boston: Harcourt Brace Jovanovich, 1987), pp. 259–82. See p. 270. Also Joseph La Dou, "The Not-So-Clean Business of Making Chips," *Technology Review*, May–June 1984, pp. 23–36, especially p. 32.

32. An introductory exposition on pubic goods may be found in Neil M. Singer, *Public Microeconomics, An Introduction to Public Finance*, 2d ed. (Boston: Little, Brown, 1972), chap. 6.

33. Stephen S. Cohen and John Zysman, *Manufacturing Matters*, p. 98.

34. Cohen and Zysman, *Manufacturing Matters*, p. 106.

35. Cohen and Zysman, *Manufacturing Matters*, pp. 220–21.

36. William G. Ouchi, *The M-Form Society* (New York: Avon Books, 1984), p. 215. To Ouchi, a successful industrial strategy of a nation would parallel that of a corporation, in that it would be organized in an M form. In an M-Form corporation, division leaders are dedicated to the competitiveness of their own productive units, while also sharing "some common endowment, such as technology, skill, or other important features" with other division heads (p. 21). Similarly, in a society having an M-Form, corporate heads would feel a responsibility both to their own firm's success and to social endowments shared with other firms.

37. Michael L. Dertouzos et al., *Made in America: Regaining the Productive Edge* (Cambridge, Mass.: MIT Press, 1989), p. 105. Still another argument about shared resources in American competitiveness is found in Bruce Scott's study, "National Strategies: Key to International Competition," in Bruce R. Scott and George C. Lodge, eds., *U.S. Competitiveness in the World Economy* (Boston, Mass.: Harvard Business School Press, 1985), pp. 71–143. He writes that a nation could accelerate industrial growth through industrial strategies. These would operate "by reshaping its resources in light of medium- or long-term market opportunities rather than by attempting to make the most efficient short-term use of existing resources." (p. 141) The resources include skill, labor, and natural resources.

38. Dertouzos et al., *Made in America*, p. 105.

FOUR

1. The Cuomo Commission on Trade and Competitiveness, *The Cuomo Commission Report* (New York: Simon and Schuster, 1988), p. 35, is one of the more recent sources that charts the decline. It cites the *U.N. Yearbook of Trade Statistics* for the export data: this adds exports from the United States, West Germany, United Kingdom, France, Japan, Canada, Italy, Holland, and Sweden.

2. One source that makes this point is Stephen S. Cohen and John Zysman, *Manufacturing Matters: The Myth of the Post-Industrial Economy* (New York: Basic Books, 1987), p. 64.

3. Cuomo Commission, *Cuomo Commission Report*, p. 186.

4. Examples of observations and recommendations of committees of notables are President's Commission on Industrial Competitiveness, *Global Competition: The New Reality*, 2 vols. (Washington, D.C.: U.S. Government Printing Office, 1985); and Business-Education Forum, "An Action Agenda for American Competitiveness," (Report by the Business-Higher Education Forum, Northwest-Midwest Congressional Coalition, and Congressional Clearinghouse for the Future, Sept. 1986). Several of the congressional hearings are cited at various points in this study.

5. Bruce R. Scott and George C. Lodge, eds., *US Competitiveness in the World Economy* (Boston, Mass.: Harvard Business School Press, 1985), especially the article by Bruce R. Scott, "National Strategies: Key to International Competition," pp. 71–143.

6. "Statement of George C. Lodge on United States Competitiveness in the World Economy," in U.S. Congress, House Subcommittee on Science, Research and Technology, *Strategies for Exploiting American Inventiveness in the World Economy* (Washington, D.C.: U.S. Government Printing Office, 1986), pp. 253–73; information cited is on p. 255.

7. Michael L. Dertouzos et al. *Made in America: Regaining the Productive Edge* (Cambridge, Mass.: MIT Press, 1989), p. 7. Earlier studies of the U.S. position in specific industries (automobile, aircraft, electronics, textiles, machine tools, pharmaceuticals, and steel) were conducted by the National Academy of Engineering and the National Research Council and are available from National Academy Press. A summary of the studies may be found in N. B. Hannay and Lowell W. Steele, "Technology and Trade: A Study of U.S. Competitiveness in Seven Industries," *Research Management*, Feb. 1986, pp. 14–22. An even earlier agency study is U.S. Department of Commerce, International Trade Administration, *An Assessment of U.S. Competitiveness in High-Technology Industries* (Washinton, D.C., 1983).

8. Quoted in Ross K. Baker, "The Bittersweet Courtship of Congressional Democrats and Industrial Policy," *Economic Development Quarterly*, 1, no. 2, (1987): pp. 111–23; citation is on p. 111.

9. R. D. Norton, "Industrial Policy and American Renewal," *Journal of Economic Literature*, Mar. 1986, pp. 1–40. The reference is on p. 34.

10. Baker, "Bittersweet Courtship."

11. For a description, see the Special Section on the Rhode Island Greenhouse Compact, *Northeast Journal of Business and Economics*, Spring–Summer 1984.

12. These were among the major reasons given by Baker, "Bittersweet Courtship."

13. Norton, "Industrial Policy and American Renewal," p. 36.

14. The diagnosis of deindustrialization, a subject that had an out-of-date flavor by the late 1980s, is best represented by Barry Bluestone and Bennett Harrison, *The Deindustrialization of America* (New York: Basic Books, 1982).

15. Robert B. Reich, "Behold! We Have an Industrial Policy," *New York Times*, May 22, 1988.

16. E. S. Savas, *Privatization: The Key to Better Government* (Chatham, N.J.: Chatham House, 1987), p. 3.

17. Ted Kolderie, "The Two Different Concepts of Privatization," *Public Administration Review*, July–August 1986, pp. 285–91. The quote is on p. 288, with stress in the original. The complete reference to "Butler" is Stuart Butler, *Privatizing Federal Spending, A Strategy to Eliminate the Deficit* (Washington, D.C.: Heritage Foundation, 1985).

18. Robert W. Bailey, "Uses and Misuses of Privatization," in Steve H. Hanke, ed., *Prospects for Privatization* (New York: Academy of Political Science, 1989), pp. 138–52.

19. Peter J. Katzenstein, *Small States in World Markets: Industrial Policy in Europe* (Ithaca, N.Y.: Cornell University Press, 1985), chap. 1.

20. Katzenstein, *Small States in World Markets*, p. 23.

21. Katzenstein, *Small States in World Markets*, p. 23.

22. John Zysman, *Government, Markets, and Growth: Financial Systems and the Politics of Industrial Change* (Ithaca, N.Y.: Cornell University Press, 1983), chap. 2.

23. Henry Ergas, "Does Technology Policy Matter?" in Bruce R. Guile and Harvey Brooks, eds., *Technology and Global Industry* (Washington, D.C.: National Academy Press, 1987), pp. 191–245. Quoted passage appears on pp. 232–33.

24. Theodore J. Lowi, *The End of Liberalism: The Second Republic of the United States*, 2d ed. (New York: W. W. Norton, 1979).

25. Lowi, *End of Liberalism*, p. 44.

26. Lowi, *End of Liberalism*, p. 56.

27. Lowi, *End of Liberalism*, p. 68.

28. Don K. Price, *The Scientific Estate* (Cambridge, Mass.: Belknap Press of Harvard University Press, 1967), chap. 2.

29. Price, *Scientific Estate*, chap. 2, especially pp. 33, 39.

30. H. L. Nieburg, *In the Name of Science* (Chicago: Quadrangle books, 1966), chap. 10. Comments on planning in the contract state appear on p. 197.

31. A recent work about industrial policy, though not by that name, is Cohen and Zysman, *Manufacturing Matters*.

32. Katzenstein, *Small States in World Markets*, p. 19.

33. Katzenstein, *Small States in World Markets*, p. 30.

34. Theodore J. Lowi, "American Business, Public Policy, Case-Studies, and Political Theory," *World Politics*, July 1964, pp. 677–715. Cited passage is on p. 690.

35. Lowi writes that this was possible in the historical transition of tariff law from distributive to regulatory policy. See Lowi, "American Business," p. 699.

36. Lowi, *End of Liberalism*, pp. 77–91.

37. In a work on science policy (Theodore J. Lowi and Benjamin Ginsberg, *Poliscide* [New York: Macmillan, 1976]), Lowi and Ginsberg suggest rule-based distributive policy. They call for planning and central supervision that take account of the spillovers that science projects have onto other social ends. They write that "this can be done by rules, laid down at the point of origin, that attempt to plan for some of the social consequences of each project" (p. 293). To do so, Congress, through the president and administrative agencies, "might exercise its wisdom rather than allow it to be displaced by local elites" (p. 294).

With respect to industrial technology, this suggestion is problematic. When we are dealing with rapid sector-specific change amid international technological and industrial rivalries, public institutions will have to respond with strategy; rules will be inadequate. Government will have to demonstrate a broader exercise of wisdom than that involved in rule making.

If this central policy-making capacity could exercise broader judgment, it would be engaging in planning. Yet Lowi and Ginsberg's own diagnosis of interest-group politics (and my present study of privatization in technology policy) serves to question a U.S. political capacity for such sustained economic judgment.

Lowi might also suggest that, by reconstituting the structure of federal budget making (through constituent policy), government could produce a more coherent technology budget. But the suggestion would require the partial removal of budgeting from the political process and the constituting of a political ability to plan. In Lowi's work, as in my present study, calls for planning pose an apparent paradox: the prescription violates the diagnoses (those of privatization or interest-group liberalism) through which it was reached. As I suggest in the concluding chapter, calls for planning may yet have a chance for viability if they build upon and reform privatized institutions.

38. Robert A. Leone and Stephen P. Bradley, "Toward an Effective Industrial Policy," *Harvard Business Review*, Nov.–Dec., 1981, pp. 91–97. Like other authors on the subject, Leone and Bradley also subscribe to other definitions. They go on to offer a managerial definition as well.

39. A listing of such federal support for sectors appears in Robert B. Reich, "Why the US Needs an Industrial Policy," *Harvard Business Review*, Jan.–Feb. 1982, pp. 74–81.

40. Aaron Wildavsky, "Squaring the Political Circle: Industrial Policy and the American Dream," in Chalmers Johnson, ed., *The Industrial Policy Debate* (San Francisco: Institute for Contemporary Studies, 1988), pp. 27–44, quote is on p. 28.

41. Lester C. Thurow, *The Zero-Sum Solution* (New York: Simon and Schuster, 1985), p. 263.

42. Reich, "Why the US Needs an Industrial Policy," p. 75. In this article, he also suggests a sectoral definition of industrial policy.

43. John Zysman and Laura Tyson, "American Industry in International Competition," in John Zysman and Laura Tyson, eds., *American Industry in International Competition* (Ithaca, N.Y.: Cornell University Press, 1983); citation is to p. 22.

44. In this genre, see James W. Botkin, Dan Dimancescu, and Ray Stata, *The Innovators, Rediscovering America's Creative Energy* (New York: Harper & Row, 1984).

45. Among the many congressional hearings that have stressed technology and innovation in competitiveness are U.S. Congress, Joint Economic Committee, *Investment in Research and Development* (Washington, D.C.: U.S. Government Printing Office, 1987); and U.S. Congress, House Subcommittee on Science, Research, and Technology, *Strategies for Exploiting American Inventiveness in the World Marketplace* (Washington, D.C.: U.S. Government Printing Office, 1986). One of the blue-ribbon committees to stress the diagnosis was the Council on Competitiveness, *Picking Up the Pace: The Commercial Challenge to American Innovation* (n.p. 1987?).

46. U.S. Department of Commerce, Office of Productivity, Technology, and Innovation, *The Importance of Productivity* (Washington, D.C., 1985?).

47. D. Bruce Merrifield, letter to Ernest Sternberg, Sept. 2, 1986.

48. Ralph Landau and Nathan Rosenberg, Introduction, to *The Positive Sum Strategy: Harnessing Technology for Economic Growth*, ed. Ralph Landau and Nathan Rosenberg (Washington, D.C.: National Academy Press, 1986), p. vi, stress in the original.

49. A good summary of this macroeconomic view is Ralph Landau, "U.S. Economic Growth," *Scientific American*, June 1988, pp. 44–52.

50. See Martin Neil Baily and Alok K. Chakrabarti, *Innovation and the Productivity Crisis* (Washington, D.C.: Brookings Institute, 1988). Also, William J. Baumol and Kenneth L. McLennan, eds., *Productivity Growth and U.S. Competitiveness* (New York: Oxford University Press, 1985).

51. This view is represented by Robert H. Hayes and Willam J. Abernathy, "Managing Our Way to Economic Decline," *Harvard Business Review*, Jul.–Aug. 1980, pp. 67–77.

52. A short treatment applied to regional development is Peter Hall, "The Geography of the Fifth Kondratieff," in Peter Hall and Ann Markusen, eds., *Silicon Landscapes* (Boston: Allen & Unwin, 1985), pp. 1–19. For more extensive treatment, see works by Christopher Freeman, including *Technology Policy and Economic Performance: Lessons from Japan* (London: Pinter, 1987).

53. See, for example, Cohen and Zysman, *Manufacturing Matters*, chap. 7.

54. David Noble, *Forces of Production* (New York: Knopf, 1984).

55. David Dickson, *The New Politics of Science* (New York: Pantheon, 1984).

56. Chalmers Johnson, "Introduction: The Idea of Industrial Policy," in Chalmers Johnson, ed., *Industrial Policy Debate* p. 9.

57. President's Commission on Industrial Competitiveness, *Global Competition*, vol. 1, p. 2.

58. For a nice summary, see Ralf Dahrendorf, *Essays in the Theory of Society* (Stanford, Calif.: Stanford University Press, 1968), chap. 8, pp. 215–31. Also see John Gray, "Hayek on the Market Economy and the Limits of State Action," in Dieter Helm, ed., *The Economic Borders of the State* (Oxford: Oxford University Press, 1989), pp. 127–43.

59. Thurow, *Zero-Sum Solution*, p. 267.

60. A recent piece to make this point is David Vogel, "Government-Industry Relations in the United States: An Overview," in Stephen Wilks and Maurice Wright, eds., *Comparative-Government Industry Relations* (Oxford: Clarendon Press, 1987), pp. 91–116.

61. See Thurow, *Zero-Sum Solution*, pp. 268–70; Reich, "Why the U.S. Needs an Industrial Policy."

62. Thurow, *Zero-Sum Solution*, p. 284.

63. Charles L. Schultze, "Industrial Policy: A Dissent," *The Brookings Review*, Fall 1983, pp. 3–12.

64. Council of Economic Advisers, *Economic Report of the President* (Washington, D.C.: U.S. Government Printing Office, 1984), chap. 3.

65. Schultze, "Industrial Policy: A Dissent," p. 11.

66. Schultze, "Industrial Policy: A Dissent," pp. 10–11.

67. Chalmers Johnson's well-known study is *MITI and the Japanese Miracle: The Growth of Industrial Policy, 1925-1975* (Stanford, Calif.: Stanford University Press, 1982).

68. Cohen and Zysman, *Manufacturing Matters*, p. 106.

69. See Zysman and Tyson, *American Industry in International Competition*, p. 33.

70. Zysman and Tyson, *American Industry in International Competition*, p. 44.

71. Schultze, "Industrial Policy: A Dissent," p. 7.

72. Schultze, "Industrial Policy: A Dissent," p. 8.

73. Council of Economic Advisers, *Economic Report*, p. 104.

74. Council of Economic Advisers, *Economic Report*, p. 106.

75. Council of Economic Advisers, *Economic Report*, p. 105.

76. Paul R. Krugman, "Targeted Industrial Policies: Theory and Evidence," in Federal Reserve Bank of Kansas City, *Industrial Change and Public Policy*, (Jackson Hole, Wyo., 1983), pp. 123–56; citation is to p. 136.

77. Gene M. Grossman, "Strategic Export Promotion: A Critique," in Paul R. Krugman, ed., *Strategic Trade Policy and the New International Economics* (Cambridge, Mass: MIT Press, 1987).

78. Thurow, *Zero-Sum Solution*, p. 293.

79. Robert Kuttner, "Commentary," in Federal Reserve Bank of Kansas City, *Industrial Change and Public Policy*, pp. 169–76; citation on p. 169.

80. Schultze, "Industrial Policy: A Dissent," p. 9.

81. It appears in Council of Economic Advisers, *Economic Report*, and repeatedly in Claude E. Barfield and William E. Schambra, eds., *The Politics of Industrial Policy* (Washington, D.C.: American Enterprise Institute for Policy Research, 1986).

82. Aaron Wildavsky, "Industrial Policies in American Cultures," in Barfield and Schambra, *The Politics of Industrial Policy*, pp. 15–32. See also Wildavsky, "Squaring the Political Circle," pp. 27–46, and Eugene Bardach, "Implementing Industrial Policy," in Johnson, ed., *Industrial Policy Debate*, pp. 91–116.

83. Thurow, *Zero-Sum Solution*, pp. 295, 296.

84. Mancur Olson, "Supply-Side Economics, Industrial Policy, and Rational Ignorance," in Barfield and Schambra, eds., *Politics of Industrial Policy*, pp. 245–69, citation is on p. 268.

85. Metaphors of individual pieces of innovation adding up to stocks and flows of technology are replete in the microeconomics of innovation. The work of Edwin Mansfield is representative. See Edwin Mansfield et al., *The Production and Application of New Industrial Technology* (New York: W. W. Norton, 1977). See also his summary article "Microeconomics of Technological Innovation," in Landau and Rosenberg, eds., *Positive Sum Strategy*, pp. 307–25.

86. Stephen J. Kline and Nathan Rosenberg, "An Overview of Innovation," in Landau and Rosenberg, eds., *Positive Sum Strategy*, pp. 275-305, citation is on p. 278.

87. Kline and Rosenberg, "An Overview of Innovation," p. 276.

88. Kline and Rosenberg, "An Overview of Innovation," p. 301.

89. An example of a prominent leader in R & D policy taking this position is Roland W. Schmidt, "National R & D Policy: An Industrial Perspective," *Science*, June 15, 1984, pp. 1206-9.

90. Nathan Rosenberg, "The Impact of Technological Innovation: A Historical View," in Landau and Rosenberg, eds., *Positive Sum Strategy*, pp. 17-32. Quoted passage appears on p. 28.

91. Giovanni Dosi, "Sources, Procedures, and Microeconomic Effects of Innovation," *Journal of Economic Literature* 26 (Sept. 1988): pp. 1120-71. Quoted passage appears on p. 1134.

FIVE

1. Disaggregation constitutes the large part of distributive policy, as that concept is used by Theodore Lowi in "American Business, Public Policy, Case-Studies, and Political Theory," *World Politics*, July 1964, pp. 677-715; the relevant passage is on p. 690. Disaggregation should be seen as distinct from the more strategic allocation of resources presided over by committees of political beneficiaries. Lowi's own work (*The End of Liberalism: The Second Republic of the United States*, 2d ed., [New York: W. W. Norton, 1979]) suggests that self-governing bodies in agriculture that preside over land reclamation and irrigation, as well as racially discriminatory urban land-use coalitions, have engaged in more coherent policy-making for their own interests. For present purposes, this collaborative form of distributive policy is seen as distincly different from its disaggregative form.

2. Public Law 100-647, a Miscellaneous Revenue Act, extended the tax credit through December 1989. Since 1988, legislation has been introduced to make the R & D tax credit permanent.

3. Executive Office of the President, Office of Management and Budget, "Special Analysis G: Tax Expenditures," in *Special Analyses, Budget of the United States Government, Fiscal Year 1988*, (Washington, D.C., 1988). Information cited is on p. G-37.

4. Questions about the value of this policy have generated a number of studies. Some of the results are reviewed in Martin Neil Baily and Alok K. Chakrabarti, *Innovation and the Productivity Crisis* (Washington, D.C.: Brookings Institution, 1988), pp. 118ff.

5. This is the Bayh-Dole Act, Public Law 96-517.

6. Public Law 98-620.

7. Public Law 99-502.

8. U.S. General Acounting Office, *Technology Transfer: Federal Agencies' Patent Licensing Activities*, GAO/RCED-91-80 (1991), pp. 6-9 and 17.

9. The $2 million figure and "postindustrial pork" term is from Lois R. Ember, "Efforts to Stem Pork-Barrel Science Funding Likely to Be Unsuccessful," *Chemical and Engineering News*, July 18, 1989, pp. 7-16; information cited is on p. 7. The latter figure is from Colleen Cordes, "Colleges Received About $289 Million in Earmarked Funds," *The Chronicle of Higher Education*, Feb. 1, 1989, p.1. An alternative enumeration is given in U.S. Congress, Office of Technology Assessment, *Federally Funded Research: Decisions for a Decade*, OTA-SET-490 (Washington D.C.: US Government Printing Office, 1991), pp. 87-93.

10. Colleen Cordes, "Colleges Received About $289 Million," p. 1.

11. Kin Ha and David Lipin, "Pork-Barreling of Science Funds" (Paper prepared under the supervision of Bruce Cain, California Institute of Technology, 1987, photocopy), p. 3.

12. Phil Kuntz, "Grants for Research Facilities: Experiment to End 'Earmarks,'" *Congressional Quarterly*, July 2, 1988, p. 1824; and Lois R. Ember, "Effort to Stem Pork-Barrel Science," p. 8.

13. Robert Rosenzweig quoted in John Walsh, "Adapting to Pork-Barrel Science," *Science*, Dec. 12, 1989, p. 1639.

14. These seemed to be the only appropriations directly related to advanced optics in a Congressional Research Service attempt to inventory earmarking for colleges: Susan H. Boren, *Appropriations Enacted for Specific Colleges and Universities by the 96th through 100th Congresses*, (Washington, D.C.: Library of Congress, Congressional Research Service, 1989).

15. John Schott, Center for Imaging Science, Rochester Institute of Technology, interview, Mar. 25, 1988.

16. Rose became president of RIT in 1979. His background is described briefly in D. R. Gordon, *Rochester Institute of Technology: Industrial Development and Educational Innovation in an American City* (New York: Edwin Mellen Press, 1982), p. 358.

17. Phil Johnston, "Shaping a Vision of Education Excellence," *Rochester Business Journal*, October 24–30, 1988, p. 10.

18. Roy Meyers, Cassidy and Associates, telephone interview, Apr. 12, 1989.

19. Marjorie Sun, "Pork Barrel Issues Simmer," *Science*, Dec. 6, 1985, p. 1145. The Congressional Research Service listing does not include any fiber-optics center at the University of South Carolina, though the university did receive other earmarked funds, such as $16.3 million for an energy research complex. The university switchboard could find no fiber-optics center in 1989.

20. Richard Stevens, Department of Energy, telephone interview, Apr. 12, 1989.

21. "Report of the advisory task force on state support for high-technology research," (Office of the Lieutenant Governor, Albany, N.Y., Jan. 1989).

22. The budget official who made the statement was interviewed by the author in May 1989.

23. The developments were covered in Patrick Corbett, "Photonics Center out of Limbo," *Rome Observer Dispatch*, Oct. 18, 1989, p. 1A; and in "National Photonics Research Center Proposed for Rome," *Technology N.Y. Report*, Mar. 1988, p. 1.

24. The amount and the speaker's response is from the budget official mentioned anonymously in the text. The Governor's response is expressed, in different terms, in the "Report of the Advisory Task Force," p. 3.

25. New York State Assembly bill #11967 in 1988.

26. "Report of the Advisory Task Force," p. 3.

27. "Report of the Advisory Task Force," p. 2.

28. The arguments are summarized in the cited articles in *Science, Congressional Quarterly*, and *Chemical and Engineering News*.

29. Quoted in "Lobbying 101: Colleges Discover the Pork Barrel," *Business Week*, Oct. 27, 1986, p. 116.

30. "Lobbying 101," p. 118; Ember, "Efforts to stem-pork barrel," p. 9. Earmarking indeed redresses the balance. Ha and Lippin's study shows that earmarked funds go overwhelmingly to "have-not" schools. The top twenty research universities received 1.3 percent of earmarked funds in 1986, while those ranking below one hundred in conventional research funding received 71 percent. However, my own purview of listings of earmarked funds shows the big-name universities appearing ever more frequently since 1986.

31. Other facilities, such as the $12 million Center for Advanced Microstructures and Devices at Louisiana State University, might work on related subjects, but one could not tell from the language in the appropriations bill. The difficulty of figuring out the purpose of earmarked facilities is inherent to this kind of legislation.

32. These examples are chosen at random from Boren, *Appropriations Enacted for Specific Colleges and Universities.*

33. Cordes, "Colleges Received About $289 Million," p. 100

34. Cordes, "Colleges Received About $289 Million," p. 100.

35. Boren, *Appropriations Enacted for Specific Colleges and Universities.*

36. Quoted in David Rapp, "Fight Over 'Academic Pork Barrel' Evaporates," *Congressional Quarterly,* Sept. 12, 1987, p. 2190.

37. John Walsh, "Adapting to Pork-Barrel Science," p. 1640.

38. "CEEE: The Center for Electronics and Electrical Engineering" (U.S. Department of Commerce, Feb. 1988).

39. Library of Congress, Congressional Research Service, *The National Bureau of Standards, A Review of Its Organization and Operations, 1971–1980,* A Study Prepared for the Subcommittee on Science, Research, and Technology, U.S. House of Representatives Committee on Science and Technology (May 1981), pp. 72, 135. Henceforth cited as CRS, *NBS, A Review.*

40. Such statements can be found, for example, in: U.S. Congress, House Committee on Science, Space, and Technology, *The Role of Science and Technology in Competitiveness,* Hearings Before the Subcommittee on Science, Research, and Technology, House of Representatives, Apr. 28, 29, and 30, 1987; and in U.S. Congress, House Committee on Science, Space and Technology, *1989 National Bureau of Standards Authorization,* Hearings Before the Subcommittee on Science, Research, and Technology, Mar. 15, 16, and 17, 1988.

41. Public Law 100–418, section 5111.

42. Aaron A. Sanders, NIST Optical Electronics Metrology Group, telephone interview, Feb. 3, 1989.

43. The preceding paragraphs reflect telephone interviews with the following persons at the National Institute of Standards and Technology: Elaine Bunten-Mines, July 25, 1991; Judson French, July 23, 1991; Robert Kamper, Apr. 26, 1989; John Lyons, Nov. 23, 1988; Aaron A. Sanders, Feb. 3, 1989; Robert I. Scace, Feb. 9, 1989; and Gregory A. Tassey, Feb. 7, 1989.

44. Gregory Tassey, *Technology and Economic Assessment of Optoelectronics*, (Gaithersburg, Md.: National Bureau of Standards, 1985).

45. French interview.

46. *Emerging Technologies in Electronics and Their Measurement Needs* (Gaithersburg, Md.: Center for Electronics and Electrical Engineering, National Engineering Laboratory, National Institute of Standards and Technology, Mar. 1989), p. 2; and *Center for Electronics and Electrical Engineering Long-Range Plan* (Gaithersburg, Md.: National Bureau of Standards, 1988), p. 13.

47. Lyons interview.

48. Tassey interview.

49. Sanders interview.

50. Testimony of Robert A. Kamper, in U.S. Congress, House Committee on Science, Space, and Technology, *Economic Development through Technology Transfer*, Hearings Before the Subcommittee on Science, Research, and Technology, Feb. 5, 1988, p. 122.

51. Scace interview.

52. These advantages of relationships with the agency are described in numerous NIST brochures and publications and were confirmed by Judson French, interview cited above.

53. For example, Kamper's testimony, hearings of Feb. 5, 1988, cited above; and testimony of Ernest Ambler, director, National Bureau of Standards, hearings of Mar. 15, 16, and 17, 1988, cited above, pp. 15–30.

54. Testimony of Lewis M. Branscomb, in U.S. Congress, House Committee on Science, Space, and Technology, *The Role of Science and Technology in Competitiveness*, Hearings Before the Subcommittee on Science, Research and Technology, Apr. 28, 29, and 30, 1987, p. 66.

55. Branscomb testimony, p. 62.

56. Branscomb testimony, p. 69.

57. Public Law 99-502

58. "LLNL to Push Laser Tech Transfer," *Photonics Spectra*, June 1991, p. 61. Not to be outdone, Los Alamos, Sandia, and other laboratories set up an Alliance for Photonics Technology with a mission "to enhance the global competitiveness of U.S. industry in the critical technology of photonics by accelerating technology transfer" to industry ("Alliance for Photoncs Technology," *Optical Materials and Engineering News*, June 1991, p. 1).

59. CRS, *NBS, A Review*, p. 70.

60. For example, Richard D. Schwartz and David S. Nelson, *The Growth and Development of the Electronic Imaging Industry* (New York: Shearson Lehman Hutton, 1987).

61. Yencharis Consulting Group, "Electronic Imaging Markets 1986-1988; Outlook, Obstacles, and Potential," cited in "Standards Key to Growth in Imaging," *Optical Engineering Reports*, Jan. 1987, pp. 1A, 3A. Another source commenting on the need for standards to preserve U.S. competitiveness in inage processing is Jawed Wahid, "Does U.S. Industry Lack the Standard Advantage?" *Chief Executive*, Iss. 52, Jul./Aug. 1989, pp. 24-27.

62. Schwartz and Nelson, *The Growth and Development of the Electronic Imaging Industry*.

63. Statement of John W. Lyons before the Subcommittee on Technology and Competitiveness, Committee on Science, Space and Technology, U.S. House of Representatives, May 21, 1991. His testimony gives more examples of NIST's involvement in imaging, but many of these were in the new Advanced Technology Program. Specifically authorized and funded by legislation when the bureau was transformed into NIST, the program provided competitively awarded industrial research grants to private firms and consortia. This new program fell outside the scope of the agency's more traditional role in metrology.

64. Lyons statement.

65. Testimony of Dr. Lawrence S. Goldberg, in U.S. Congress, House Committee on Science, Space, and Technology, Subcommittee on Investigation and Oversight, *The Cutting Edge: Science and Technology for America's Future* (Washington, D.C.: U.S. Government Printing Office, 1988), pp. 145-50.

66. Proceedings are summarized in *The Future of Lightwave Technology: An NSF Workshop* ([1984]).

67. Goldberg testimony, p. 149.

68. Nam P. Suh, Massachusetts Institute of Technology, telephone interview, Feb. 16, 1989. Suh also makes such points in a memorandum to the author dated July 18, 1991.

69. Suh interview.

70. Suh interview.

71. Documents produced by the task forces included: National Science Foundation Task Force for Strategic Planning, "A strategic plan for the Engineering Directorate of the National Science Foundation," Nov. 28, 1984; and National Science Foundation, Task Group on Directorate for Engineering Reorganization Plan, "Report on Directorate for Engineering reorganization plan," internal NSF memorandum, Dec. 7, 1984.

72. Suh interview.

73. Paul J. Herer, National Science Foundation, telephone interview, Feb. 21, 1989.

74. Estimates obtained through telephone interviews with Albert Harvey, program director, Lightwave Technology Program, Jan. 25, 1989; and Allen R. Stubberud, University of California, Irvine, formerly division director, NSF Division of Electrical, Communications, and Systems Engineering, Sept. 30 1988.

75. "Peer Review in Selected Federal Agencies," in National Science Foundation Advisory Committee on Merit Review, *Final Report* (Washington, D.C.: National Science Foundation, 1986), pp. A1–A16.

76. National Science Foundation Advisory Committee on Merit Review, *Final Report*, p. 10.

77. National Science Foundation Advisory Committee on Merit Review, *Final Report*, p. 10

78. Proposal Evaluation Form, National Science Foundation Form 1, Apr. 1984.

79. "Peer Review in Selected Federal Agencies," p. A4.

80. National Science Foundation Advisory Committee on Merit Review, *Final Report*, p. 12.

81. Harvey Brooks, "The Problem of Research Priorities," *Daedalus* 107, no. 2, (Spring 1978): pp. 171–90, cited in U.S. Congress, Office of Technology Assessment, *Federally Funded Research: Decisions for a Decade*, p. 147. Also see discussion and references in chap. 5 of the same report.

82. National Science Foundation Advisory Committee on Merit Review, *Final Report*, p. 12.

83. Harvey interview.

84. "Program Announcement, Lightwave Technology," National Science Foundation Directorate for Engineering [1987].

85. Lawrence Goldberg, program director, Quantum Electronics, Waves, and Beams Program, National Science Foundation, telephone interview, Feb. 1, 1989.

SIX

1. David Noble, *America by Design* (Oxford: Oxford University Press, 1977), p. 122.

2. Carmela Hacklisch, Herbert I. Fusfeld, and Alan D. Levenson, *Trends in Collective Industrial Research*, 2d ed. (New York: New York University Graduate School of Business Administration, Center for Science and Technology Policy, May 1986), pp. 32, 38–41.

3. Richard Corigan, "Antitrust Culture Shock Dazes Firms Freed to Join Hands in R & D Ventures," *National Journal*, Nov. 17, 1984. The act is Public Law 98–462.

4. U.S. Department of Commerce, Office of Productivity, Technology and Innovation, "R & D Ventures Reported Under the National Cooperative Research Act," Annual listings for 1985, 1986, and 1987.

5. The beginnings of an avalanche of literature on the subject include Lansing Felker, "Cooperative Industrial R&D: Funding the Innovation Gap," *Bell Atlantic Quarterly*, Winter 1984; and Andrew Pollack, "Uniting to Create Products," *New York Times*, Jan. 14, 1986, p. D1.

6. U.S. Department of Commerce, "R & D Ventures" lists Bellcore-Honeywell, Bellcore-Heinrich Hertz Institut, Bellcore-Hitachi, Bell Communications-Triquent, Bell Communications-Microwave Services Corp., and Corning Glass-Nippon Telegraph and Telephone each as a "consortium" working in optoelectronics, or fiber optics.

7. Robert Holman, telephone interviews, Apr. 27, 1989 and Aug. 2, 1991.

8. Mark B. Myers, vice president, Xerox-Webster Research Center, interview, Feb. 4, 1988. The quotation includes minor changes suggested by Myers in a telephone conversation with the author on Aug. 29, 1991.

9. Kathleen M. Buyers and David R. Palmer, "The Microelectronics and Computer Technology Corporation: An Assessment from Market and Public Policy Perspectives," *Administration and Society* 21, no. 1, (May 1989): pp. 101–27. Citation is to p. 125.

10. Most of these smaller, institutionally created research centers are not further discussed here. One can gather that there are dozens by browsing through "Optics Education, 1989" (SPIE—International Society for Optical Engineering, Bellingham, Wash., 1989); or by counting programs indexed under "optics" and related terms in the *Research Centers Directory*, 13th ed., 1989.

11. Just a small selection includes Government-University-Industry Research Roundtable, *New Alliances and Partnerships in American Science and Engineering* (Washington, D.C.: National Academy Press, 1986); U.S. Department of Commerce, *The New Climate for Joint Research* (Washington, D.C., 1984); Thomas W. Lanffitt et al., eds., *Partners in the Research Enterprise* (Philadelphia, Pa.: University of Pennsylvania Press, 1983); *Journal of the Society of Research Administrators* 17, no. 2, (Fall 1985), special issue on university-industry joint research; *Corporate and Campus Cooperation: An Action Agenda* (Washington, D.C.: Business-Higher Education Forum, 1984); *Converting U.S. Scientific Leadership to Technological Leadership*, Proceedings of the 1984 Conference on Industrial Science and Technological Innovation, sponsored by the National Science Foundation and hosted by North Carolina State University, May 14–16, 1984 (Raleigh, N.C., [1984?]).

12. Remarks on the acceleration of technological change (as a reason for participation in collaborative arrangements) were prominent in interviews that this author conducted with Edwin P. Przybylowicz, senior vice president for research, Eastman Kodak Co., by phone, Feb. 9, 1988; and Mark B. Meyers, vice president, Xerox-Webster Research Center, Feb. 4, 1988 and Nov. 4, 1988.

13. Government-University-Industry Research Roundtable, *New Alliances and Partnerships*, p. ix.

14. Noble, *America by Design*, p. 136.

15. Government-University-Industry Research Roundtable, *New Alliances and Partnerships*, p. 1.

16. Several acts of policy-making suggest this intent. For example, revisions of the Patent and Trademark Act (Bayh-Dole Act, Public Law 96–517) allowed universities to retain patents issued under federally funded research and, thereby, to license them to corporate participants in campus research. Moreover, the intent was reflected, as we shall see below, in the actions of several state science agencies, the National Science Foundation, and other federal agencies to establish collaborative university-industry research centers.

17. References to such advantages are scattered throughout the many proceedings of conferences and collections of papers, cited above, on joint research centers

18. K. R. Swartzel and D. O. Gray, "Industry-University Research in Agriculture and Food Science," *Food Technology*, Dec. 1987.

19. David A. Tansik, "University of Arizona Optical Circuitry Cooperative," Report prepared for the National Science Foundation (University of Arizona College of Business and Public Administration, Tucson, Ariz. 1985, unpaginated).

20. Claude Barfield, "Out of the Laboratory into the World Market," *Government Executive*, Jan. 1988, pp. 22–26; and Panel on Science and Technology Centers, *Science and Technology Centers, Principles and Guidelines* (Washington, D.C.: National Academy of Science, 1987), p. 19; and National Science Foundation, "NSF Science and Technology Centers," Apr. 1991.

21. Testimony of Dr. Lawrence S. Goldberg, in U.S. Congress, House Committee on Science, Space and Technology, Subcommittee on Investigations and Oversight, *The Cutting Edge: Science and Technology for America's Future* (Washington, D.C.: U.S. Government Printing Office, 1988), pp. 145–150.

22. Syl McNinch, *Engineering Research Centers (ERC): How They Happened, Their Purposes and Comments on Related Programs* (Washington, D.C.: National Science Foundation, 1984), pp. 1–3.

23. McNinch, pp. 4–5; and *Engineering: An Expanded and More Active Role for NSF* (Washington, D.C.: National Science Foundation, 1985).

24. *Guidelines for Engineering Research Centers* (Washington, D.C.: National Academy of Engineering, 1983). Despite the year indicated on the report's cover, it was apparently submitted in February 1984.

25. McNinch, p. 11; and *Engineering: An Expanded and More Active*, p. 2.

26. Letter from Robert M. White, National Academy of Engineering, to Edward A. Knapp, National Science Foundation, Feb. 15, 1984. Another example comes from the NSF's first Engineering Research Centers announcement, prepared in 1984. Calling for university proposals to set up engineering research centers, it stated as the first of a list of features that the centers would "Provide research opportunities to develop fundamental knowledge in areas critical to US competitiveness in world markets."

27. Robert M. White letter to Edward A. Knapp

28. *Strengthening Engineering in the National Science Foundation* (Washington, D.C.: National Academy of Engineering, 1983), p. 5.

29. "Division of Cross-Disciplinary Research," (National Science Foundation Directorate of Engineering, Washington, D.C. Oct. 18, 1988, Photocopy) p. 2.

30. U.S. General Accounting Office, *Engineering Research Centers: NSF Program Management and Industry Sponsorship*, (1988).

31. USGAO, *Engineering Research Centers*, p. 20.

32. Program Announcement, Engineering Research Centers, Fiscal Year 1988, National Science Foundation Directorate for Engineering, Washington, D.C..

33. Lynn Preston, deputy director, Division of Cross-Disciplinary Research, telephone interview, Feb. 21, 1989.

34. Program Announcement, Engineering Research Centers, 1989, National Science Foundation Directorate for Engineering, Washington, D.C., p. 2.

35. *State Technology Programs in the United States* (St. Paul, Minn.: Office of Science and Technology, Minnesota Department of Trade and Economic Development), 1988, p. 6.

36. *State Technology Programs in the United States*, p. 97.

37. "Report of the advisory task force on state support for high-technology research," (Office of the Lieutenant Governor, Albany, N.Y., Jan. 1989), p. 30.

38. Kitta Macpherson, " 'Photonics' Center for New Jersey," *Newark Star Ledger*, Apr. 21, 1988, pp. 1ff.

39. James E. Underwood, *Science/Technology Related Activities in the Government of New York State: The Organizational Pattern* (Albany, N.Y.: Office of Science and Technology, New York State Education Department, 1971).

40. John Rozett, New York State Assembly, interview, May 23, 1988.

41. New York State Laws of 1982, Chapter 562, Sec. 1.

42. Batelle Memorial Institute, Columbus Laboratories *Development of High Technology Industries in New York State, Final Summary Report Prepared for the New York State Science and Technology Foundation* (Columbus, Ohio, 1982).

43. New York State Science and Technology Foundation, *Directory of Projects Fiscal Years 1980–1989: A Decade of Progress* (Albany, N.Y., 1990), p. 3.

44. New York State Science and Technology Foundation, *Annual Report, 1983–84* (Albany, N.Y., [1984]) p. 2.

45. Kitta MacPherson, " 'Photonics' Center for New Jersey."

46. Vernon Ozarow, New York State Science and Technology Foundation, interview, Aug. 27, 1986.

47. An example of such a report is *New York State Center for Advanced Optical Technology, The Institute of Optics* (Rochester, N.Y.: University of Rochester, 1988).

48. The description of the Institute of Optics reflects interviews with Duncan T. Moore, Dec. 10, 1987; Bruce Arden, Mar. 14, 1988; Don Hess, Mar. 14, 1988; and Dennis Hall, April 13, 1989, all of the University of Rochester.

49. *Center for Telecommunications Research, Industrial Participants Program* (New York: Columbia University, School of Engineering and Applied Science, [1986]); and The New York State Center for Advanced Optical Technology, *Prospectus for sponsors* (Rochester, N.Y.: University of Rochester, College of Engineering and Applied Science, [1987]).

50. USGAO, *Engineering Research Centers*, p. 38.

51. Robert Shannon, director, Optical Sciences Center, University of Arizona at Tucson, telephone interview, Apr. 18, 1989.

52. Robert Guenther, Army Research Office, Research Triangle Park, N.C., telephone interview, Dec. 8, 1988.

53. Robert H. Leshne, "The American Precision Optics Manufacturing Association (APOMA)," ([1989], Manuscript).

54. Paul Christianson, Columbia University, Telecommunications Research Center, interview, Sept. 3, 1986.

55. Frank A. Darknell and Edith C. Darknell, "State College Science and Engineering Faculty's Collaborative Links with Private Business and Industry in California and Other States," in National Science Board, *University-Industry Research Relationships: Selected Studies* (Washington, D.C.: National Science Foundation, [1982]), pp. 165–92; reference is to p. 177.

56. Robert Shannon, director, Optical Sciences Center, University of Arizona at Tucson, telephone interview, Apr. 17, 1989. Shannon adds in a telephone interview on July 31, 1991, that he does not intend to be cynical. "There are sincere university efforts to integrate industry and university research," he says. "But if one is to be successful in raising research funds, one has to follow the pattern that the government is establishing."

57. Thomas Cathey, director, Center for Optoelectronic Computing Systems, University of Colorado at Boulder, telephone interview, Apr. 17, 1989.

58. George H. Sigel, director, Fiber Optics Materials Research Program, Rutgers University, telephone interview, Apr. 18, 1989.

59. Richard Perloff, "Case Western Reserve University Center for Applied Polymer Research," in Division of Industrial Science and Technological Innovation, *Development of University-Industry Cooperative Research Centers: Historical Profiles* (Washington D.C.: National Science Foundation, [1984?]), pp. 19–24, quotation is on p. 20.

60. Arnost Reiser, director, Institute of Imaging Science, Polytechnic University, telephone interview, Apr. 19, 1989.

61. The information is from a selection of brochures on RIT Research Corporation.

62. Hess interview.

63. Sigel interview.

64. For an extensive discussion of legal issues in university-industry research arrangements, including sample texts of contracts, see Bernard D. Reams, *University-Industry Research Partnerships* (Westport, Conn.: Quorum, 1986).

65. Tansik, "University of Arizona Optical Circuitry Cooperative" (unpaginated).

66. For examples of this concern outside the optics world, see Division of Industrial Science and Technological Innovation, *Development of University-Industry Cooperative Research Centers.*

67. U.S. Patent Office, Office of Technology Assessment and Forecast, *TAF Report Top U.S. Universities*, and *TAF Report—Lightwave Technology* ([Washington, D.C.:] U.S. Patent and Trademark Office, 1987, Text-fiche). Another source of 1986 patent data shows university patenting rates that are about 40 percent higher, but the discrepancy does not alter my conclusion that university patenting rates are relatively low. See National Science Board, *Science & Engineering Indicators—1989* (Washington, D.C.: Government Printing Office, 1989).

68. U.S. Patent Office, *TAF Reports—Top US Universities* and *TAF Report—Lightwave Technology*.

69. Hall interview.

70. Christianson interview.

71. Cathey interview.

72. Such points were made by Christianson, cited above; Sigel, cited above; and Joseph Verdeyen, director, Center for Compound Semiconductor Microelectronics, University of Illinois at Urbana-Champaign, telephone interview, Apr. 26, 1989, among others.

73. Hall interview. In his interview, Shannon at the University of Arizona, takes a similar view. Dennis Hall adds in an interview on August 9, 1991, that industrial research representatives are well meaning, but the blending of educational and industrial missions can be problematic.

74. William F. Hamilton, "Corporate Strategies for Managing Emerging Technologies," *Technology in Society* 7, nos. 2-3, (1985): pp. 197-212.

75. Statement made in confidence by a staff member of the New York State Science and Technology Foundation.

76. Lynn Preston, "Engineering Research Centers Program, Review System Guidelines" (National Science Foundation, Division of Cross-Disciplinary Research, Washington, D.C., 1988), p.3.

77. Center for Telecommunications Research, *Overview and Strategic Plan* (New York: Columbia University School of Engineering and Applied Science, 1989), p. 14.

78. Center for Compound Semiconductor Microelectronics, University of Illinois at Urbana-Champaign, College of Engineering, *Strategic Plan* (Urbana, Ill. 1987).

79. Government-University-Industry Research Roundtable, *New Alliances and Partnerships*, p. 102.

80. Howard Moraff, Division of Cross-Disciplinary Research, National Science Foundation, telephone interview, May 5, 1989.

81. Lynn Preston, National Science Foundation, telephone interview, July 23, 1991.

82. Verdeyen interview.

83. Christianson interview.

84. Paul Christianson, Columbia University Telecommunications Research Center, telephone interview, July 24, 1991.

85. Interviews with Sigel, Verdeyen, Cathey, and others, cited above.

SEVEN

1. John Couretas, "Defense Marketers Get a Jump on New Business with IR&D," *Business Marketing*, July 1985, pp. 42–48.

2. Library of Congress, Congressional Research Service, *Science Support by the Department of Defense*, Transmitted to the Task Force on Science Policy, Committee on Science and Technology, U.S. House of Representatives (Washington, D.C.: U.S. Government Printing Office, 1987), pp. 321–57. Henceforth cited as *Science Support*.

3. *Science Support*, pp. 305–10.

4. *Science Support*, pp. 301–5.

5. A description of the program appears in Congressional Budget Office, *The Benefits and Risks of Federal Funding for Sematech* (1987).

6. *Bolstering Defense Industrial Competitiveness*, Report to the Secretary of Defense by the Under Secretary of Defense (Acquisition) ([Washington, D.C.:] Department of Defense, 1988), p. iii.

7. *Bolstering Defense Industrial Competitiveness*, p. 5.

8. U.S. General Accounting Office, *Industrial Base: Defense Critical Industries*, GAO/NSIAD-88-192BR (Washington, D.C.: 1988), pp. 24–28, lists 185 of the industries.

9. U.S. Congress, Office of Technology Assessment, *The Defense Technology Base: Introduction and Overview*, OTA-ISC-374 (Washington, D.C.: U.S. Government Printing Office, 1988), p.7. Henceforth cited as OTA, *Defense Technology Base*.

10. OTA, *Defense Technology Base*, p. 7.

11. U.S. Congress, Office of Technology Assessment, *Holding the Edge: Maintaining the Defense Technology Base*, OTA–ISC–420 (Washington, D.C.: U.S. Government Printing Office, 1989), p. 161 n. 1. Henceforth cited as OTA, *Holding the Edge*.

12. Defense Science Board, *The Defense Industrial and Technology Base*, vol. 1 of the 1988 Summer Study (Washington, D.C.: Office of the Under Secretary of Defense for Acquisition, 1988), p. 2. Henceforth cited as 1988 Summer Study.

13. *Bolstering Defense Industrial Competitiveness*, p. 1.

14. *Report of the Secretary of Defense Frank C. Carlucci to the Congress on the FY 1990/FY 1991 Biennial Budget and FY 1990–1994 Defense Programs* (Washington, D.C.: U.S. Government Printing Office, 1989), p. 119.

15. Air Force Association and the USNI Military Database, *Lifeline in Danger: An Assessment of the United States Defense Industrial Base* (Arlington, Va.: Aerospace Education Foundation, 1988).

16. Ad Hoc Industry Advisory Committee, "Report to the Subcommittee on Defense Industry and Technology, Senate Armed Services Committee," [Under the letterhead of John D. Rittenhouse, Senior V.P., General Electric Company, Cherry Hill, N.J.,] February 5, 1988.

17. U.S. General Accounting Office, *Industrial Base*.

18. OTA, *Holding the Edge*.

19. "Background," DoD's undated information circular on the Defense Science Board, lists the Board's summer studies through 1988.

20. 1988 Summer Study, vol. 1, p. 55.

21. The nine tasks appear in the original in bulleted form. Only four are reproduced here. See the 1988 Summer Study, vol. 2, pp. A2–A3.

22. *Bolstering Defense Industrial Competitiveness*, p. 11; stress is in the original.

23. *Report of the Secretary of Defense Frank C. Carlucci*, p. 119.

24. Department of Defense, *Critical Technologies Plan*, For the Committees on Armed Services, United States Congress, March 15, 1989. The statutory wording requiring the plan is given in appendix B. Henceforth cited as *Critical Technologies Plan*.

25. However, according to the *Critical Technologies Plan*, p. 2, the plan omits technologies related to nuclear weapons.

26. *Critical Technologies Plan*, p. A-47.

27. "U.S. demonstrates advanced weapons technology in Libya," *Aviation Week and Space Technology*, Apr. 21, 1986, pp. 18-21.

28. Bruce R. Mackie, "Modern Military: Imaging Is Everything," *Laser Focus*, Aug. 1988, p. 148.

29. Malcolme W. Browne, "Invention That Shaped the Gulf War: The Laser-Guided Bomb," *New York Times*, Feb. 26, 1991, pp. C1, C8.

30. The commentary was unanimous on this point. See, for example, Richard W. Stevenson, "In the Gulf, a 'Black Box' War," *New York Times*, Jan. 20, 1991, sec. 3, pp. 1, 6; John D. Morrocco, "Gulf War Boosts Prospects for High-Technology Weapons," *Aviation Week & Space Technology*, Mar. 18, 1991, pp. 45-47, and Craig Couvault, "Desert Storm Reinforces Military Space Directions," *Aviation Week & Space Technology*, Apr. 8, 1991, pp. 42-47.

31. "Integrated Systems Design Addresses 21st Century Combat Environment," *Aviation Week & Space Technology*, Apr. 14, 1986, pp. 69-71.

32. James R. Kanely and Lawrence R. Kilty, "The Government Market for Fiberoptic Systems," *Laser Focus*, Apr. 1984, pp. 93-108.

33. Kanely and Kilty, "The Government Market for Fiberoptic Systems;" and John Rhea, "Fiber Optics Expands Design Potential for Weapons Systems." *Laser Focus*, Aug. 1987, pp. 94-99.

34. "Fiber Optics: The Big Move in Communications and Beyond," *Business Week*, May 21, 1984, p. 172; and *Critical Technologies Plan*, p. A-33.

35. More extensive description of the military uses of fiber optics appears in C. David Chaffee, *The Rewiring of America: The Fiber Optics Revolution* (Boston: Academic Press, 1988), chap. 11.

36. Other compound semiconductor materials are indium phosphide, indium antimonide, and mercury cadmium telluride.

37. *Critical Technologies Plan*, p. A-9.

38. *Critical Technologies Plan*, p. A-29.

39. Critical Technologies Plan, p. A-29.

40. In the commercially irrelevant pulsed power that energizes laser weapons, "The United States is the undisputed free-world leader," the *Critical Technologies Plan* observed. "There are few significant commercial applications" in technologies of signature control, a field in which "the allies are several years behind the U.S." In most types of phased arrays, which have only military application, this country has a substantial lead. Also "there is no commercial market for sensitive passive arrays," a field in which the U.S. has the largest body of technical literature. And laser radars have little commercial value. (*Critical Technologies Plan*, pp. A–75, A–62, A–55, A–47, A–39).

41. *Critical Technologies Plan*, p. A–35.

42. *Critical Technologies Plan*, p. A–31.

43. *Critical Technologies Plan*, p. A–11.

44. Clayton Jones, "Japanese High-Tech in the Gulf," *Christian Science Monitor*, Feb. 7, 1991, p. 4.

45. See p. 94 in Michael Brody, "The Real World Promise of Star Wars," *Fortune*, June 23, 1986, pp. 92–98.

46. A review of the issues surrounding SDI as industrial policy appears in Walter Zegveld and Christian Engzing, *SDI and Industrial Technology Policy: Threat or Opportunity* (New York: St. Martin's Press, 1987).

47. David Kales, "SDI Makes Opportunities for Laser and Electro-optics Businesses" [Interview with James A. Ionson], *Laser Focus*, Mar. 1988, pp. 86–88. Quote is on p. 86. Ionson uses the term "electro-optics" broadly to include computing and high-energy lasers.

48. "Oversell Hurts Optical Computing," *SDI Monitor*, Feb. 2, 1987, p. 40.

49. See Kevin L. Zonderman, "Optical Computers: A Decisive Edge for SDI," *Beam Technology Report: Fusion*, Mar.–Apr. 1986, pp. 25ff.

50. William J. Broad, *Star Warriors* (New York: Simon and Schuster), 1985, p. 60.

51. Jeff Hecht, "Optical Computers," *High Technology*, Feb. 1987, pp. 44–49; and H. John Caulfield, "Optical Computing: Some Hard Questions," *Optics News*, Apr. 1986, p. 10.

52. Trudy E. Bell, "Optical Computing: A Field in Flux," *IEEE Spectrum*, Aug. 1986, pp. 34–57. Citation is to p. 55.

53. The company, Guiltech Research, is described in Hecht, "Optical Computers," p. 49.

54. Bell, "Optical Computing," p. 49.

55. Bell, "Optical Computing," p. 56.

56. Kevin L. Zondervan, "Optical Computers: A Decisive Edge for SDI."

57. Alexander Wolfe, "Optical Computing is Beginning to Take on the Glow of Reality," *Electronics Week*, June 10, 1985, pp. 24–27. Quote appears on p. 26.

58. Caulfield is quoted to that effect in Rudy Abramson, "Supporting Spinoffs Will Help Foot Star Wars Research Bill," *Roanoke Times and World News*, Jan. 4, 1987, p. A8. "My greatest concern," he said, "is that this country will lose optics to the Japanese the way we lost electronics."

59. Malcolm W. Browne, "The Star Wars Spinoffs," *New York Times Magazine*, Aug. 24, 1986, pp. 22–29. Subtitle appears on p. 22.

60. Browne, "The Star Wars Spinoffs," p. 66.

61. "Pushing Computers to the Ultimate Speed Limit," *Business Week*, July 28, 1986, pp. 48–49. The quote appears on p. 48.

62. "Pushing Computers to the Ultimate Speed Limit," p. 49.

63. H. John Caulfield, "Optical Computing: Some Hard Questions," *Optics News*, Apr. 1986, p.10. Similar remarks appear in "Oversell Hurts Optical Computing," p. 40.

64. A chronology of developments appears in Jun Shibata and Takao Kajiwara, "Optics and Electronics are Living Together," *IEEE Spectrum*, Feb. 1989, pp. 34–38.

65. Shibata and Kujiwata, "Optics and Electronics are Living Together;" and *Critical Technologies Plan*, p. A–11.

66. Zegveld and Enzing, *SDI and Industrial Technology Policy*, p. 99.

67. *Science Support*, p. 309.

68. *Science Support*, p. 305.

69. Frost and Sullivan Inc., *The Military Gallium Arsenide Semiconductor Market in the U.S.* (1987), p. 185.

70. Defense Science Board, *Science and Technology Base Management*, 1987 Summer Study (Washington, D.C.: Office of the Secretary of Defense, 1987), p. 47. Henceforth cited as 1987 Summer Study.

71. "MIMIC Winners Get to Work," *Electronics*, June 1988, pp. 37, 40.

72. Quoted in Tim Carrington, "Military Dependence on Foreign Suppliers Causing Rising Concern," *Wall Street Journal*, Mar. 24, 1988.

73. Institute of Electrical and Electronics Engineers, "Report of the Task Force on Defense Electronic Materials" [Report delivered at a conference in Arlington, Va., ?], Apr. 9, 1985. Since the report consists of a set of bulleted items reproduced from viewgraphs, the logic must be inferred.

74. Institute of Electronics and Electrical Engineers, "Briefing Paper on the Strategic Materials Initiative," Washington, D.C.: [1987?], p. 2. The quoted passage is underlined in the original.

75. Institute of Electronics and Electrical Engineers, "Briefing Paper on the Strategic Materials Initiative," p. 2.

76. Arati Prabhakar, director, Microelectronic Technology Office, DARPA, telephone interview, Aug. 13, 1991.

77. For a list of the studies, see U.S. General Accounting Office, *Industrial Base*, pp. 41–42.

78. Robert H. Leshne, "The American Precision Optics Manufacturers Association (APOMA)" (Draft paper, [Jan. 1989?]), p. 1. Henceforth referred to as APOMA.

79. Statement by Robert Leshne in response to questioning by Senator Bingaman, in U.S. Congress, Senate Committee on Armed Services, *Manufacturing Capabilities in Key Second-Tier Industries*, Hearings Before the Senate Subcommittee on Defense Industry and Technology, Senate Hearing 100–512, July 23, 1987.

80. APOMA, p. 3.

81. Joint Group on the Industrial Base, Joint Precision Optics Technical Group, "Joint Logistics Commanders Precision Optics Study," Wright-Patterson AFB, Ohio, June 19, 1987, pp. 1, 9. Henceforth cited as "JLC Precision Optics Study."

82. "JLC Precision Optics Study," pp. 2–4, 50–60.

83. "JLC Precision Optics Study," pp. 5, 25.

84. Statement of Dr. Paul Freedenberg, Under Secretary of Commerce for Export Administration, in Senate Hearing 100–790, pt. 7, p. 192.

85. Freedenberg statement p. 192.

86. Freedenberg statement p. 197.

87. Quoted in Tim Carrington, "Military's Dependence on Foreign Suppliers Causes Rising Concern," *Wall Street Journal*, Mar. 24, 1988.

88. David Lytle, "Optics and DoD: Who Are We Defending?" *Photonics Spectra*, May 1988, pp. 48–49.

89. Lytle, "Optics and DoD."

90. Carrington, "Military's Dependence on Foreign Suppliers."

91. Lytle, "Optics and DoD" p. 48.

92. "Prepared Statement by Dr. R. H. Leshne," in Senate Hearing 100–790, pt. 7, p. 217.

93. Phil Ebersole, "Funding Set for New Optics Center," *Democrat and Chronicle* (Rochester, N.Y.), July 14, 1989, p. 8D.

94. APOMA, p. 1.

95. APOMA, p. 3.

96. Carrington, "Military's Dependence on Foreign Suppliers." Membership in the Precision Optics Manufacturing Association is given as forty-eight in Gary M. Kaye, "Building an Optics Consortium," *Photonics Spectra*, Aug. 1989, p. 38.

97. David Silverberg, "Foreign Optics Dependence Concerns Six Senators," *Defense News*, Jan. 30, 1989, p. 31.

98. Letter to Alfonse D'Amato from Robert C. McCormack, Office of the Secretary of Defense, Feb. 7, 1989.

99. Letter from Congressmen Horton, Bennett, Nelson, Ireland, La Falce, Udall, Kolbe, and Slaughter to Darold Griffin, Army Material [sic] Command, May 26, 1989.

100. Ebersole, "Funding Set for New Optics Center."

101. Memorandum from Robert C. McCormack to assistant secretaries of the army, navy, and air force and directors of the Defense Logistics Agency and the Strategic Defense Initiative Organization, Nov. 6, 1989.

102. Sandra Sugawara, "Pentagon Plans to Aid Four Ailing Industries," *Washington Post*, Jan. 2, 1990, sec. D, pp. 13, 15.

103. "APOMA Jubilant over Army Contract," *Photonics Spectra*, Aug. 1990, p. 42.

104. Background information on the agency may be found in *Science Support*, pp. 101–8; and OTA, *Holding the Edge*, pp. 52–56.

105. Edward Dolnick, "DARPA: The Pentagon's Skunk Works," *Across the Board*, Apr. 1984, pp. 28–35; cited information is on p. 30. Other sources with such lists are *Science Support*, p. 105; and Jonathan Jacky, "The Strategic Computing Program," in David Bellin and Gary Chapman, eds., *Computers in Battle: Will They Work?* (Boston: Harcourt Brace Jovanovich, 1987), pp. 171–337; cited information is on p. 172.

106. *The Defense Advanced Research Projects Agency (DARPA)* (Washington, D.C.: Department of Defense, [1987]), p. 20. Henceforth cited as *DARPA prospectus*. An extensive account of the controversy is found in Jacky, "Strategic Computing Program."

107. See Andrew Pollack, "Pentagon Sought Smart Truck but It Found Something Else," *New York Times*, May 30, 1989, p. 1. See also Jacky, "Strategic Computing Program."

108. John Neff, former program officer at DARPA, telephone interview, June 8, 1989.

109. Neff interview.

110. Andrew Yang, DARPA, telephone interview, July 21, 1989.

111. "Optical Computing Enhancement Project Gets DARPA Grant," *OE Reports*, Nov. 1989, p. 5.

112. *DARPA Prospectus*, p. 11.

113. Articles presenting a good review of the technical and economic issues appear in *Technology Review*, Apr. 1989. Numerous articles have appeared in the *New York Times, Wall Street Journal*, and *Business Week*. A few of them, along with a few of the congressional hearings, are cited below.

114. Proposals of the American Electronics Association especially highlighted the interindustry effects. See "Summary of the AEA Response on ATV to the Requests of the Subcommittee on Telecommunications and Finance of the Committee on Energy and Commerce of the U.S. House of Representatives," in *Public Policy Implications of Advanced Television Systems*, A Staff Report Prepared for the Use of the Subcommittee on Telecommunications and Finance, U.S. House of Representatives, Committee Print 101-E, Mar. 1989, pp. 20–45. Another report to emphasize multiple industrial effects is U.S. Department of Commerce, National Telecommunications and Information Administration, *Advanced*

Television, Related Technologies, and the National Interest, (n.p., 1989). "Super Television," a special section of *Business Week*, Jan. 30, 1989, pp. 56–66, is more accessible.

115. Statement of Robert A. Mosbacher before the Subcommittee on Telecommunications and Finance, Committee on Energy and Finance, U.S. House of Representatives, Mar. 8, 1989.

116. Both quotes and the description of the Mosbacher initiative appear in "Mosbacher's Initiative on HDTV Is Getting Scuttled, Sources Say," *Wall Street Journal*, Aug. 2, 1989, p. 82.

117. Statement by Craig I. Fields, deputy director for research, Defense Advanced Research Projects Agency, before the House Subcommittee on Telecommunications and Finance, Mar. 8, 1989 (unpaginated); and "The US's Semiconductor Battle Plan," *Broadcasting*, Dec. 26, 1988, p. 33.

118. "Super Television," p. 63.

119. Quoted in "Defense Department Wants in the HDTV Picture," *Broadcasting*, Dec. 26, 1988, pp. 32–33.

120. Fields statement.

121. Fields statement.

122. Fields statement.

123. "Defense Department Wants in the HDTV Picture," p. 33; and Patterson interview.

124. Patterson interview; also "Defense Department Wants in the HDTV Picture," p. 32.

125. Patterson interview.

126. Fields statement.

127. "Television and the Economy: An Interview with William F. Schreiber," *Technology Review*, Apr. 1989, p. 17.

128. Thomas G. Donlan, "Redoubtable DARPA," *Barron's*, Apr. 3, 1989, pp. 14–15.

129. Fields statement.

130. Fields statement.

131. Fields elaborated in his testimony that "we are now reviewing the proposals and opportunities [presented by the responses to the high

definition display initiative]. We will form these teams into a cohesive program and that will be done very soon. For example in the supercomputer part of our Strategic Computing initiative we arranged for a number of computer companies to cooperate in the use of chip making facilities that they could not have afforded on their own."

132. Bob Davis, "Ouster of Defense Aide Craig Fields Sparks Discord, Congressional Criticism," *Wall Street Journal*, Apr. 23, 1990, p. A6.

133. "The Government's Guiding Hand: An Interview with Ex-DARPA Director Craig Fields," *Technology Review*, Feb.–Mar. 1991, pp. 35–40, esp. p. 36.

134. Davis, "Ouster of Defense Aide."

135. John Burgess and Stuart Auerbach, "Official's Tansfer Prompts Hill Inquiry," *Washington Post*, Apr. 25, 1990, pp. G1, G4.

136. "Dreams of Fields," *Wall Street Journal*, Apr. 25, 1990, p. A14.

137. "Government's Guiding Hand," p. 36.

138. "News from Jeff Bingaman, U.S. Senator," News release dated May 3, 1991.

139. "Optoelectronics and All-Optic Networks," part of an attachment to a letter from Charles M. Herzfeld, director of Defense Research and Engineering, to Sam Nunn, chairman, Committee on Armed Services, U.S. Senate, Apr. [12 or 18?], 1991.

140. Prabhakar interview.

141. "DARPA and the DOD" [Interview with Donald Atwood], *Defense Electronics*, July 1991, pp. 32–33.

142. "DARPA Decides on New Chief," *Optics and Photonics News*, Feb. 1991, p. 29.

143. Andrew Pollack, "America's Answer to Japan's MITI," *New York Times*, Mar. 5, 1989, business section pp. 1, 8. Citations are to p. 8.

144. Fields statement.

145. Graham's criteria are "first, the state of a particular technological area; second, the effectiveness of the competing approaches; third, the leverage of the technology on the overall system performance; and fourth, the availability and readiness of competent technical teams." Statement of Dr. William Graham, in U.S. Congress, Senate, *Hearings on Defense Industry and Technology*, Senate Hearing 100–242, pt. 7, (1987), p. 3245.

146. Freedenberg statement, p. 195.

147. Examples may be found in the Statement of Dr. Ronald Kerber, in Senate Hearing 100–242, pt. 7, pp. 3269–70; and in 1988 Summer Study, vol. 2, p. 11.

148. *Critical Technologies Plan*, p. 5.

149. 1988 Summer Study, p. 10.

150. OTA, *Holding the Edge*, p. 20.

151. John D. Morocco, "Air Force Altering Budget Priorities for Project Forecast 2," *Aviation Week & Space Technology*, Apr. 13, 1987, p. 22.

152. Phillip Speser, "Air Force Emphasizes Laser and Electro-optics Research," *Laser Focus*, May 1988, p. 20.

153. "Project Forecast II: Executive Summary," Air Force Systems Command, Andrews AFB; and Morocco, "Air Force Altering Budget," p. 22.

154. Morocco, "Air Force Altering Budget," p. 22.

155. David Hugh, "Griffiss Lab Leads USAF Drive to Tap Potential of Photonics," *Aviation Week & Space Technology*, Jan. 30, 1989, p. 54.

156. See the list of sixteen studies in the 1987 Summer Study, p. 58. The study itself is the seventeeth.

157. OTA, *Holding the Edge*, p. 47.

158. Statement of Dr. Robert B. Costello, in Senate Hearing 100–790, pt. 7, p. 187.

159. *Bolstering Defense Industrial Competitiveness*, p. 64.

160. *Report of the Secretary of Defense Frank C. Carlucci*, p. 120.

161. Ed McGaffigan, office of Senator Bingaman, telephone interview, July 18, 1989.

162. Military policies toward makers of weapons and other military items would qualify by this definition but would offer a relatively uninteresting example for present purposes. The policies are more interesting when applied to industries in which military acquisitions consume only a small part of the output or to technologies that have significantly more than military uses.

163. Senate Hearing 100-242, p. 3348.

164. McGaffigan interview.

EIGHT

1. "Swan Song for Laissez-faire?" *Business Week*, Special Issue on Innovation in America, 1989 [no month given], p. 174. The idea is promoted by a lobbying group, Rebuild America, supported both by the American Electronics Association, a strong backer of federal initiatives toward high-definition TV, and by the National Center for Manufacturing Sciences, a consortium supported in large part by the Defense Department.

2. Practices of the National Institute of Standards of Technology, a civilian agency, to choose fields of research on technical measurement did represent quite wide-ranging thinking about technological priorities, but these efforts were susceptible to the vagaries of business participation and the budget process.

3. The early 1990s revealed indications of new developments in U.S. industrial technology policy. The Bush adminstration's Office of Science and Technology Policy issued a formal "technology policy" document, one that emphasized private sector leadership. The first years of the 1990s also saw the issuing of several critical technologies lists for civilian industry, in forms similar to those prepared for military use in the 1980s. And concepts of "generic" and "precompetitive" technology gained currency as formulas for justifying technology policy, though it was unclear whether these ideas differed from that of "public goods" for understanding technology. It remained to be seen whether these developments would significantly redirect the trend toward privatized industrial policy that had evolved in the 1980s.

4. On biotechnology, a good overview is U.S. Congress, Office of Technology Assessment, *New Developments in Biotechnology: U.S. Investment in Biotechnology—Special Report*, OTA-BA-360 (Washington, D.C.: U.S. Government Printing Office, 1988). On government involvement in the evolution of computer technology, see Kenneth Flamm, *Targeting the Computer: Government Support and International Competition* (Washington, D.C.: Brookings Institution, 1987).

5. See R. Scott Fosler and Renee A. Berger, *Public-Private Partnerships in American Cities* (Lexington, Mass.: Lexington Books, 1982); and Perry Davis, ed., *Public-Private Partnerships: Improving Urban Life* (New York: Academy of Political Science, 1986).

6. See Donald C. Baumer and Carl E. Van Horn, *The Politics of Unemployment* (Washington, D.C.: Congressional Quarterly, 1985); "Industry/education partnerships," *VocEd*, Special Issue, May 1983; and Elizabeth L. Useem, *Low Tech Education in a High Tech World: Corporations and Classrooms in the New Information Society* (New York: Free Press, 1986).

7. For example, see Lowi and Ginsberg's call for planning in technology policy in Theodore J. Lowi and Benjamin Ginsberg, *Poliscide* (New York: Macmillan, 1976), chap. 13.

8. Kenneth Flamm has already made a proposal along such lines. He proposes a national technology office to monitor U.S. technological capability relative to other nations and report observations for public debate. At the same time, the office would "serve as a custodian of the public interest in those associations formed to carry out joint industrial research." See Flamm, *Targeting the Computer*, p. 198.

9. In 1971, author Bruce L. R. Smith expressed this dilemma in his introduction to a volume of papers on public accountability in research contracting, ("Accountability and Independence in the Contract State," in Bruce L. R. Smith and D. C. Hague, *The Dilemma of Accountability in Modern Government* [New York: St. Martin's Press, 1971], p. 4.). He wrote:

The problem can be broadly stated at the start as the need to create understandings, and institutional arrangements, that will enable government (in both its legislative and executive branches) to maintain a strong central policy direction over the apparatus of private institutions performing services for the government while giving the private institutions enough independence of operation to produce the maximum incentives for a distinctive and creative contribution to the government.

"There is," Smith added, "to some extent an inescapable tension between the values of independence and accountability" (p. 5).

Select Bibliography

Ausubel, Jesse H., and Dale H. Langford. *Lasers, Invention to Application*. Washington, D.C.: National Academy Press, 1987.

Baily, Martin Neil, and Alok K. Chakrabarti. *Innovation and the Productivity Crisis*. Washington, D.C.: Brookings Institute, 1988.

Barfield, Claude E., and William E. Schambra, eds. *The Politics of Industrial Policy*. Washington, D.C.: American Enterprise Institute for Policy Research, 1986.

Bernstein, Jeremy. *Three Degrees Above Zero*. New York: Charles Scribner's Sons, 1984.

Bolstering Defense Industrial Competitiveness. Report to the Secretary of Defense by the Under Secretary of Defense (Acquisition). [Washington, D.C.,] Department of Defense, 1988.

Broad, William J. *Star Warriors*. New York: Simon and Schuster, 1985.

Bromberg, Joan Lisa. *The Laser in America, 1950–1970*. Cambridge, Mass.: MIT Press, 1991.

Burrows, William E. *Deep Black: Space Espionage and National Security*. New York: Random House, 1986.

Buyers, Kathleen M., and David R. Palmer, "The Microelectronics and Computer Technology Corporation: An Assessment from Market and Public Policy Perspectives." *Administration and Society* 21, no. 1, (May 1989): pp. 101–27.

Chaffe, C. David. *The Rewiring of America: The Fiber Optics Revolution.* Boston: Academic Press, 1988.

Cohen, Stephen S., and John Zysman. *Manufacturing Matters, The Myth of the Post-Industrial Economy.* New York: Basic Books, 1987.

Council of Economic Advisers. *Economic Report of the President.* Washington, D.C.: U.S. Government Printing Office, 1984.

Cuomo Commission on Trade and Competitiveness. *The Cuomo Commission Report.* New York: Simon and Schuster, 1988.

Dahrendorf, Ralf. *Essays in the Theory of Society.* Stanford, Calif.: Stanford University Press, 1968.

Defense Science Board. *The Defense Industrial and Technology Base* vol. 1 of the 1988 Summer Study. Washington, D.C.: Office of the Under Secretary of Defense for Acquisition, 1988.

Dertouzos, Michael L., Richard K. Lester, Robert M. Solow, and the MIT Commission on Industrial Productivity. *Made in America: Regaining the Productive Edge.* Cambridge, Mass.: The MIT Press, 1989.

Dosi, Giovanni, "Sources, Procedures, and Microeconomic Effects of Innovation." *Journal of Economic Literature.* 26 (Sept. 1988): pp. 1120–71.

Dosi, Giovanni, Christopher Freeman, Richard Nelson, Gerald Silverberg, and Luc Soete, eds. *Technical Change and Economic Theory.* London: Pinter, 1988.

Ergas, Henry. "Does Technology Policy Matter?" *Technology and Global Industry*, edited by Bruce R. Guile and Harvey Brooks. Washington, D.C.: National Academy Press, 1987.

Faltas, Sami. "The Invention of Fibre-Optic Communications." *History of Technology* 5 (1988): pp. 31–49.

Federal Reserve Bank of Kansas City. *Industrial Change and Public Policy.* Jackson Hole, Wyo., 1983.

Fishlock, David, ed. *A Guide to the Laser.* New York: American Elsevier, 1967.

Flamm, Kenneth. *Targeting the Computer: Government Support and International Competition.* Washington, D.C.: Brookings Institution, 1987.

Forman, Paul. "Behind Quantum Electronics: National Security as Basis for Physical Research in the United States, 1940–1960." *Historical Studies in Physical and Biological Sciences* 18, no. 1 (1987): pp. 150–229.

Freeman, C., J. Clark, and L. L. G. Soete. *Unemployment and Technical Innovation.* London: Pinter, 1982.

Freeman, Christopher. *Technology Policy and Economic Performance: Lessons from Japan.* London: Pinter, 1987.

Government-University-Industry Research Roundtable. *New Alliances and Partnerships in American Science and Engineering.* Washington, D.C.: National Academy Press, 1986.

Gray, John. "Hayek on the Market Economy and the Limits of State Action." *The Economic Borders of the State*, edited by Dieter Helm, pp. 127–43. Oxford: Oxford University Press, 1989.

Hacklisch, Carmela, Herbert I. Fusfeld, and Alan D. Levenson. *Trends in Collective Industrial Research*, 2d ed. New York: New York University Graduate School of Business Administration, Center for Science and Technology Policy, 1986.

Hall, Peter, and Ann Markusen, eds. *Silicon Landscapes.* Boston: Allen & Unwin, 1985.

Hamilton, William F. "Corporate Strategies for Managing Emerging Technologies." *Technology in Society* 7, nos. 2–3 (1985): pp. 197–212.

Hughes, Thomas P. *Networks of Power: Electrification in Western Society, 1880–1930.* Baltimore: Johns Hopkins University Press, 1983.

Johnson, Chalmers, ed. *The Industrial Policy Debate.* San Francisco: Institute for Contemporary Studies, 1988.

———. *MITI and the Japanese Miracle: The Growth of Industrial Policy, 1925–1975.* Stanford, Calif.: Stanford University Press, 1982.

Joint Group on the Industrial Base, Joint Precision Optics Technical Group. "Joint Logistics Commanders Precision Optics Study." Wright-Patterson AFB, Ohio.

Katzenstein, Peter J. *Small States in World Markets: Industrial Policy in Europe.* Ithaca, N.Y.: Cornell University Press, 1985.

Kingslake, Hilda G. *The Institute of Optics, 1929–1987.* Rochester, N.Y.: University of Rochester College of Engineering and Applied Science, 1987.

Kolderie, Ted. "The Two Different Conceps of Privatization." *Public Administration Review* July–Aug. 1986, pp. 285–91.

Krugman, Paul R., ed. *Strategic Trade Policy and the New International Economics.* Cambridge, Mass.: MIT Press, 1987.

Kuhn, Thomas S. *The Structure of Scientific Revolutions*, 2d edition. Chicago: University of Chicago Press, 1970.

Lakoff, Sanford, and Herbert F. York. *A Shield in Space? Technology, Politics, and the Strategic Defense Initiative.* Berkeley and Los Angeles: University of California Press, 1989.

Library of Congress. Congressional Research Service. *Science Support by the Department of Defense.* Transmitted to the House Task Force on Science Policy, Committee on Science and Technology. Washington, D.C.: U.S. Government Printing Office, 1987.

Lowi, Theodore J. "American Business, Public Policy, Case-Studies, and Political Theory." *World Politics*, July 1964, pp. 677–715.

———. *The End of Liberalism: The Second Republic of the United States*. 2d ed. New York: W. W. Norton, 1979.

Lowi, Theodore J. and Benjamin Ginsberg. *Poliscide*. New York: Macmillan, 1976.

National Research Council. *Photonics: Maintaining Competitiveness in the Information Era*. Washington, D.C.: National Academy Press, 1988.

Nieburg, H. L. *In the Name of Science*. Chicago: Quadrangle Books, 1966.

Noble, David. *America by Design*. Oxford: Oxford University Press, 1977

———. *Forces of Production*. New York: Knopf, 1984.

Norton, R. D. "Industrial Policy and American Renewal." *Journal of Economic Literature*, Mar. 1986, pp. 1–40.

Optoelectronic Industry and Technology Development Association, *Annual Report 1987*. (Tokyo, 1988)

———. *Annual Report 1988*. (Tokyo, 1989)

Ouchi, William G. *The M-Form Society*. New York: Avon Books, 1984.

President's Commission on Industrial Competitiveness. *Global Competition: The New Reality*, 2 vols, Washington, D.C.: U.S. Government Printing Office, 1985.

Price, Don K. *The Scientific Estate*. Cambridge, Mass.: The Belknap Press of Harvard University Press, 1967.

Reams, Bernard D. *University-Industry Research Partnerships*. Westport, Conn.: Quorum, 1986.

Savas, E. S. *Privatization: The Key to Better Government*. Chatham, N.J.: Chatham House, 1987.

Schultze, Charles L. "Industrial Policy: A Dissent," *The Brookings Review*, Fall 1983, pp. 3–12.

Schumpeter, Joseph A. *Capitalism, Socialism, and Democracy.* 1942. Reprint. New York: Harper and Row, 1975.

Schwartz, Richard D., and David S. Nelson. *The Growth and Development of the Electronic Imaging Industry: Opportunities and Challenges.* New York: Shearson Lehman Hutton, 1987.

Science Applications International Corporation. *JTech Panel Report on Opto and Microelectronics.* US Dept. of Commerce Contract No. TA-83-SAC-02254. Springfield, Va.: National Technical Information Service, 1985.

Scott, Bruce R., and George C. Lodge, eds. *U.S. Competitiveness in the World Economy.* Boston, Mass.: Harvard Business School Press, 1985.

Seidel, Robert. "From Glow to Flow: A History of Military Laser Research and Development," *Historical Studies in Physical and Biological Sciences* 18, no. 1, (1987), pp. 104–47.

Senior, J. M. and Ray, T. E. "Optical-Fibre Communications: The Formation of Technology Strategies in the UK and USA," *International Journal of Technology Management* 5, no. 1 (1990): pp. 71–88.

Sharp, Margaret E., ed. *Europe and the New Technologies.* Ithaca, N.Y.: Cornell University Press, 1986.

Smith, Bruce L. R., and D. C. Hague. *The Dilemma of Accountability in Modern Government.* New York: St. Martin's Press, 1971.

Sobel, Michael I. *Light.* Chicago: University of Chicago Press, 1987.

Su, Frederick, ed. *Technology of Our Times: People and Innovation in Optics and Optoelectronics.* Bellingham, Wash.: SPIE Press, 1990.

Tassey, Gregory. *Technology and Economic Assessment of Optoelectronics*. Gaithersburg, Md.: National Bureau of Standards, 1985.

Thompson, Brian J. "Education in optics—challenges at hand." In *Proceedings of the Society of Photo-Optical Instrumentation Engineers*, vol. 978. Bellingham, Wash.: 1989.

U.K. Advisory Council on Science and Technology. *Optoelectronics: Building on Our Investment*. London: Her Majesty's Stationery Office, 1988.

U.S. Congress. Office of Technology Assessment. *The Defense Technology Base: Introduction and Overview*. OTA-ISC-374, Washington, D.C.: U.S. Government Printing Office, 1988.

———. *Federally Funded Research: Decisions for a Decade*. OTA-SET-490, Washington, D.C.: U.S. Government Printing Office, 1991.

———. *Holding the Edge: Maintaining the Defense Technology Base*. OTA-ISC-420. Washington, D.C., U.S. Government Printing Office, 1989.

U.S. Congress. Senate. Committee on Commerce, Science, and Transportation. *Laser Technology, Development, and Applications*. Hearings before the Subcommittee on Science, Technology and Space. Serial No. 96-106, Washington, D.C.: U.S. Government Printing Office, 1980.

U.S. General Accounting Office. *Technology Transfer: Federal Agencies' Patent Licensing Activities*. GAO/RCED-91-80, 1991.

U.S. International Trade Commission. *U.S. Global Competitiveness: Optical Fibers, Technology and Equipment*. USITC Publication 2054. Washington, D.C., 1988.

Wilks, Stephen, and Maurice Wright, eds. *Comparative-Government Industry Relations*. Oxford: Clarendon Press, 1987.

Zysman, John. *Government, Markets, and Growth: Financial Systems and the Politics of Industrial Change.* Ithaca, N.Y.: Cornell University Press, 1983.

Zysman, John, and Laura Tyson, eds. *American Industry in International Competition.* Ithaca, N.Y.: Cornell University Press, 1983.

INDEX

Accountability, 289 n.9
Advisory Council on Science and Technology (U.K.), 32, 34–35, 247 n.69
Air Force Forecast II, 219–220
American Electronics Association, 288 n.1
American Precision Optics Manufacturers Association, 162, 201–206, 283 n.96
Army Materiel Command, 201–202, 205–206
Association of American Universities, 117
AT&T, 25, 27, 28, 126, 131, 157, 195, 214, 216
Atwood, Donald, 216
Augustine, Norman, 183
Austria, 82

Ball bearing industry, 203
Bardach, Eugene, 101
Battelle Columbus Laboratories, 143, 156
Battelle Optoelectronics Group, 143
Bausch and Lomb, 160
Bell Communications Research, 157
Bell Laboratories. See AT&T
Bingaman, Jeff, 216, 225
Biotechnology, 33, 39, 80, 234, 248 n.70
Black agencies, 29
Boeing Co., 216
Boskin, Michael, 215
Branscomb, Lewis, 128

Broad, William J., 194
Bromberg, Joan Lisa, 26
Brown, George, 152
Bush (presidential) administration, 211, 215, 288 n.3
Business participation
at National Institute of Standards and Technology, 128–131
in merit review, 134–138

Capitalism. See Political economy of capitalism
Carey (gubernatorial) administration, 156
Carlucci, Frank, 182, 184
Carter (presidential) administration, 76, 115
Case Western Reserve University, 165
Cassidy and Associates, 118, 123
Caulfield, John H., 196, 197
Centers of excellence, 148
Cohen, Stephen, 69–70, 97
Cold war
military-industrial policy after, 177
power structure of, 85–86, 230
reconnaissance in aftermath of, 30
Collaboration (see also Industry-university joint research), 12–13, 141–174, 231
as framework for workable industrial policy, 237
compared to business participation in government labs, 12–13

Collaboration *continued*
  compared to distributive policy, 141
  industrial consortia, 142–144
  reasons for, 271 n.12
Colorado Advanced Technology Institute, 155
Colorado State University, 151
Columbia University, 145, 157, 160, 162–163, 168, 170, 172
Common resources, 45–46, 63–65, 88
  and policy-making knowledge, 69
  as argument for planning, 65
  as justification for intervention, 65–69
  attainments of skill as, 64–65, 69–70, 88
  coherent allocation of, 46
  compared to collective endowments, 70–71
  compared to factors of production, 70
  compared to public goods, 67–71
  in M-Form corporations, 255 n.36
  in national industrial advantage, 69–71, 255 n.37
  irrigation systems as, 64
  kinds of, 58–59, 63–65
  policy research on, 234
  public infrastructure as, 65, 88
  river valleys as, 64
  U.S. inability to make policies toward, 58–59
Compound semiconductors (*see also* Optoelectronic circuits), 19
  environmental effects of, 66
  in military applications, 190–191
  in U.S. Department of Defense, 197–200

Contract state, 84–86
Cornell University, 119, 171
Corning Inc., 27, 131, 160
Corporatism, 82–83, 86–87
Costello, Robert, 180–181, 203
Creative destruction, 49, 55–57
Criteria (for industrial or technology policy), 11, 58–61, 224, 286 n.145
  and forecasting, 219–221
  and technology lists, 218–219
  at DARPA, 212–215
  in military agencies, 217–218
Critical technology lists, classifications in, 219
Critical Technologies Plan (U.S. Dept. of Defense), 185–186, 191–192, 219
Cuomo, Mario, 120

D'Amato, Alfonse, 118, 205
DARPA (Defense Advanced Research Projects Agency) (*see also* Fields, Craig), 177–178, 179, 207–217, 232
  and display technology, 210–215
  and industrial consortia, 209, 285 n.131
  and optical computing research, 208–209
  as originator of technologies, 207–208
  compared to MITI, 207, 210, 214–215
  decision making in, 209, 213–215
  disbursal of pork-barrel funding, 122
  expenditures on opto-electronics research, 40
  involvement in civilian industry, 213, 216, 223–224
  missions of, 207, 223–224
  organization of, 208

DARPA *continued*
  origins of, 24, 207
  program in optoelectronic
    materials, 198–200
  setting of research agendas at,
    209–210, 213–215
  strategic optoelectronics
    materials initiative, 198–200
  Strategic Computing Program,
    208
Defense Advanced Research
  Projects Agency. *See* DARPA
Defense industrial base, 180–182
Defense Science Board, 182–183, 198, 219
Defense technology base, 181
Democratic party politics, 232
Dickson, David, 93
Disaggregation (*see also* Merit review, Pork-barrel appropriations), 11–12, 113–139, 231
  definition of, 113
  compared to distributive policy, 265 n.1
  in policies toward photonics, 138–139
  types of, 114
Discreteness, technological
  compared to integrality, 102, 229
  in microeconomics, 103–104
Display technology, 21, 210–215
  and fiber optics, 242 n.18
  applications of, 211, 212
  at DARPA, 210–215
Distributive policy, 87–88
  compared to disaggregation, 87, 265 n.1
Domenici, Pete, 122, 199
Dosi, Giovanni, 48–50, 107
Dual-use technology, 181–182, 221

Earmarking. *See* Pork barrel appropriations
Eastman Kodak Co., 23, 32, 57, 120, 157, 160, 209, 253 n.25, 271 n.12

Economic Report of the President, 97
Electrification, 54
Electro-optics, 186–188
  usage of the term, 241 n.4
  in the Strategic Defense Initiative, 194
Electronic Industries Association, 125, 127
Engineering Research Centers, 149–155
  guidelines for, 170, 171–172
  origins of, 151–153
Ergas, Henry, 83–84
Europe
  in display technology, 191
  performance in photonics, 39
Externalities, 65–67

Factors of production, 70, 87
Federal Acquisition Regulation, 201–206
Federal Laboratory Consortium for Technology Transfer, 115
Federalism by contract, 85
Feldstein, Martin, 96–100
Fiber optics, 18–19, 26–28
  and optoelectronic translation, 18
  at Rutgers University, 164
  communications, 18, 28
  encryption, 189
  firms in, 31
  measurement research on, 125–126
  medical applications of, 66
  military research on, 28
  military technology, 189–190
  origins of, 26–28
  patenting rates in, 37–38
  U.S. trade performance in, 36
Fields, Craig, 212–215, 217
  Congressional testimony of, 212–213
Filière, 253 n. 23

Flamm, Kenneth, 289 n.8
Forecasting of technological change, 219-221
Forman, Paul, 30
France, 27, 37-38, 83
Freedenberg, Paul, 203, 218
Freeman, Christopher, 49-50

Gallium arsenide, 19, 66, 179, 190, 198, 219
Gansler, Jacques, 225
Gazelle Microcircuits Inc., 215
General Electric Co., 209, 216
Generic technology, 288 n.3
Germany (See West Germany)
Government-University-Industry Research Roundtable, 147-148, 171
Graham, William, 218
Greenhouse Compact, 77
Grossman, Gene, 99
Gulf War, 188
  military technology in, 188, 244 n.43
  US. reliance on Japanese technology, 192

Hamilton, William, 169
Hayek, Friedrich, 94-95
High definition television. See Display technology
High-technology industries, 91, 92
Holman, Robert, 143
Huang, Alan, 195
Hughes Aircraft Company, 25, 183, 216
Hughes, Thomas, 54
Human resource policy, 64-65, 80-81, 88

IBM, 195, 216
Ideology, 110, 225
Imaging, 20-22, 29-31
  as systems of devices, 21-22
  by satellites, 29-31
  firms involved in, 31-32
  image-capture devices, 20
  military applications of, 186-188
  National Institute of Standards & Technology, 129-131
  origins of, 29
  Rochester Institute of Technology, 118-119, 121
  uses of, 22
Independent Research and Development (IR&D), 178
Industrial consortia, 142-144, 270 n.6
Industrial decline (U.S.), 74-76
  and interest rates, 92
  and management practices, 75, 92
  and the inability to plan, 6
  and the reconceiving of the economy, 180-182
  blue-ribbon panels on, 182-184, 256 n.4
  commom resources and, 69-71
  defense industrial base, 180-182
  defense technology base, 181
  high-technology industries, 75
  in optoelectronics, 216
  in photonics, 36-39, 229
  military concerns about, 175-177, 180, 182-183
  precision optics, 202
  privatization as explanation for, 14, 233
  trade deficit, 74-75
Industrial policy (see also Military industrial policy), 9, 73-111
  definition of, 9-10, 88-94, 109
  administrative structures of, 236-237
  and high-technology industries, 91

Industrial policy *continued*
  Bush administration, 211
  compared to technology policy, 91–94
  criteria for, 11, 59–61, 99–100, 254 n.28
  criticisms of, 9
  "crowd" in Congress, 215
  de facto, 89, 95, 96
  debate over, 9–10, 77–79, 86–88, 94–101, 110
  euphemisms for, 9, 78, 227
  government's incapacity in, 100–101
  in Japan, 41–42
  institutional framework for, 66–67, 235, 236
  politics of, 76–77
  rejection of, 73–74
  sublimation of, 78
  tensions to be resolved in, 237
Industrial Policy Committee (proposed), 183–184
Industrial policy proposals (U.S.)
  failure of, 77
  origins of, 73
Industrial sectors
  in definition of industrial policy, 90–91
  in military industrial policy, 203–205, 224–225
Industry-led policy, 227, 288 n.1
Industry-university joint research (*see also* Research agendas), 144–173
  absence of policy research in, 171
  accumulated designations in, 162–163
  as centers of technological strength, 148, 158
  as fund raising mechanism, 162–163, 275 n.56
  as privatized industrial policy, 173–174
  as technology policy, 173
  corporate funding for, 161–163
  growth of, in photonics, 52
  in the states, 155–159
  legal issues in, 275 n.64
  organization of, 160–163
  origins of, 147–148
  patenting in, 166–167
  philanthropic motives for, 169
  rationales for, 146–149
  recruiting of bus. affiliates, 160–162
  research centers in photonics, 145–146, 271 n.10
  scholarly accomplishment and, 166
  why businesses participate, 168–169
  "Window strategies" of corporations in, 169
Infant industries, 92, 198
Innovation
  as assemblage of discrete inventions, 45
  "black box" of, 104–105
  in competitiveness, 102
  microeconomics of, 103–104
Institute of Electrical and Electronics Engineers, 198–199
Integrality
  and knowledge of technology, 59, 71
  as argument for industrial policy, 108
  as justification for public intervention, 58–61, 65–66
  compared to discreteness, 102, 229
  compared to linkages, 108
  compared to side-effects as reason for intervention, 65–67
  compared to systems, 54
  in technological change, 10, 105–108

Integrality *continued*
  market's inefficiency in
    response to, 59–61
  policy research on, 233–234,
    254 n.28
  U.S. policy response to, 71
Interest rates, 92
Interest-group liberalism, 84–85,
  88
International Society for Optical
  Engineering, 23, 161
Ionson, James A., 194, 196–197

Japan (*see also* MITI,
    Optoelectronic Industry and
    Technology Development
    Association), 27, 33–42, 75,
    83
  corporate involvement in
    photonics, 32
  Gulf War, 192
  in display technology, 191
  industrial policy, 41–42
  institutional framework of its
    industrial policy, 236
  military applications of
    optoelectronics, 191–192
  optoelectronics production, 32,
    33, 36–39, 197
  photonics R&D, 4, 39–41
  rates of publishing in
    optoelectronics, 38
Johnson, Chalmers, 42, 93, 96
Joint Logistical Commanders'
  Precision Optics Study,
  201–204, 223
JTech (Japanese Technology)
  Panel, 38

Katzenstein, Peter, 82–84,
  86–87, 94
Kennedy, Edward, 76, 119
Keyworth, George, 151
Kline, Stephen, 104–105
Knapp, Edward, 151

Knowledge, 10
  about technological paradigms,
    52–53
  and common resources, 69
  as made possible by
    integrality, 59–60
  as theme in industrial policy
    debate, 10, 102
  contextual (in markets), 94,
    105, 106
  in definition of industrial
    policy, 89, 90–91
  in industrial policy making,
    107–108, 109, 224–225
  in the industrial policy debate,
    97–100, 107–108
  synoptic (in the state), 94
  tacit, 47
Kolderie, Ted, 79–80
Krugman, Paul, 99
Kuhn, Thomas, 46–47
Kuttner, Robert, 100

Landau, Ralph, 92
Lasers
  entrepreneurship in, 26
  invention of, 23–25, 106
  military compared to
    commercial value of, 280
    n.40
  military research on, 24–26,
    244 nn.41, 42
  military uses of, 25
  "solution in search of a
    problem," 25
  U.S. trade performance in,
    36–37
  uses of, 17, 106
Lawrence Livermore National
  Laboratory, 129, 268 n.58
Learning, as argument for
  industrial policy, 97, 98
Leshne, Robert, 201, 204–205
Liberal capitalism, 81–84
Libya, U.S. raid on, 188

Lightwave technology
  usage of the term, 241 n.4
  National Science Foundation,
    131–134
  patenting rates, 37–38, 248
    n.79
Lincoln Laboratories
  (Massachusetts Institute of
  Technology), 200
Linkages, 96–97, 98, 108
Lobbying, by optics firms, 205
Lockheed Corp., 182, 183, 209
Lodge, George L., 75
Low, George M., 152
Lowi, Theodore J., 84–88, 101,
  230, 235, 258 n.37, 265 n.1
Lyons, John, 127, 268 n.63

Maiman, Theodore H., 25
Manufacturing Technology (Man
  Tech) Program, 179, 198
Market failure, 97, 99
Market-led capitalism, 81–84
Massachusetts Institute of
  Technology, 147, 167
Maurer, Robert D., 27
Merit review, 134–138, 232–233
  Advisory Committee on,
    135–136
  as changed from peer review,
    135–136
  for Engineering Research
    Centers, 153–155
  procedures of, 134–135, 153–154
Merrifield, D. Bruce, 91
Microelectronics and Computer
  Technology Corp., 144, 209,
  215
Microwave, Millimeter, and
  Monolithic Integrated Circuit
  (MIMIC) Program, 198
Military industrial policy, 13,
  222–226
  definition of, 222–223, 287 n.162
  policy-making knowledge in,
    224–225

Military industrial programs (see
  also DARPA, Precision
  Optics, Strategic Defense
  Initiative)
  as a clutter of discrete choices,
    224–225
  as privatized industrial policy,
    225–226
  concerns about indus. decline,
    176–178
  coordination of, 182–184,
    221–222
  industrial databases, 221–222
  in photonics, 177–178, 192
  planning of, 217–222
  political coalitions supporting,
    177, 216–217, 225–226,
    232
  proliferation of, 176–177
  rationales for, 176–177,
    212–213
Military technologies
  compound semiconductor
    materials, 190–191
  critical technologies lists of,
    185–186
  display technology, 210–214
  electro-optics, 186–188
  fiber optics, 189–190
  imaging, 186–188
  in U.S. raids and incursions,
    188
  precision optics, 201–202
  role of information in, 185
  U.S. compared to Japan,
    191–192
  U.S. compared to Soviet Union,
    191
Miller, Mel, 120
MITI (Ministry of International
  Trade and Industry, Japan),
  41–42, 95, 96, 236
  compared to DARPA, 207, 210,
    214–215
Mobilization in war, 175

Mondale, Walter, 76–77
Moore, Duncan, 204
Mosbacher, Robert A., 211

National Academy of
 Engineering, 133, 151–152
National Bureau of Standards.
 See National Institute of
 Standards and Technology
National Cooperative Research
 Act, 142, 233
National industrial advantage
 common resources in, 255 n.37
 political economy of, 234
 role of paradigms in, 63
 role of public policy in, 110
National Institute of Standards
 and Technology, 39,
 123–131, 138, 232
 Advanced Technology
  Program, 268 n.63
 business participation in,
  128–131
 imaging research, 129–131
 research in physical
  measurement, 124–125
 role in standard-setting, 124
 setting of research priorities
  at, 127–131, 288 n.2
National investment bank, 77
National Reconnaissance Office,
 29–30
National Research Council, 4,
 36–37, 157
 report on photonics, 4
National Science Foundation
 (*see also* Engineering
 Research Centers, Merit
 review), 131–139
 and industry-university joint
  research, 144, 147–159
 Engineering Directorate
  reorganization, 131–134,
  149–153
 grant review procedures in,
  134–138

National Science Board, 151, 154
 setting of research agendas in,
  136–138
National Security Agency, 189
National Technical Information
 Service, 115
National Technology Office
 (proposed), 289 n.8
Neff, John, 208–209
New Jersey Commission on
 Science and Technology,
 155, 157–158
New York State Science and
 Technology Foundation,
 156–159, 163, 169, 170, 233
Nieburg, N.L., 84–86, 230
Noble, David, 93, 147
Nunn, Sam, 122, 199, 225

Office of Technology Assessment
 (U.S. Congress), 42, 62, 181
Olson, Mancur, 102, 108
Optical computing, 19
 at DARPA, 208–209
 controversy over its feasibility,
  195–197
 in the Strategic Defense
  Initiative, 192–197
 military applications, 190–191
Optical Society of America, 23,
 161
Optics (*See also* Photonics,
 Precision optics), 16
 history of, 22–23
Optoelectronic circuits (*see also*
 Compound semiconductors),
 190–191, 194, 197–200
Optoelectronic Industry and
 Technology Development
 Association (Japan), 32, 33,
 246 n.68, 247 n.69
Optoelectronics (*See also*
 Photonics)
 usage of the word ,240 n.4
 effects on industrial sectors,
  34–35

Optoelectronics *continued*
  effects on photographic
    industry, 57, 253 n.25
  in computing, 19
  market shares, 33, 36–39
  market value of, 34–35
  National Institute of Standards
    and Technology, 125–127
  production in Japan, 3
  rates of publishing on, 38
  research expenditure in,
    39–41, 249 n.83, 250 n.89
  U.S. industrial decline in, 216
Ouchi, William, 70, 255 n.36
Outline (of this book), 8–14

Paradigm. *See* Technological
  paradigm, Scientific
  paradigms
Partnership, ethic of, 70, 227
Patents
  federal licensing to private
    firms, 115
  in lightwave technology, 37–38
  in university research,
    166–167, 272 n.16
Peer review. *See* Merit review
Photonics (*see also* Fiber optics,
  Electo-optics, Imaging,
  Lasers, Lightwave
  technology, Optical
  Computing, Optics,
  Optoelectronics, Precision
  Optics), 1–4, 16–17,
  228–230
  definition, 16, 240 n.4
  applications of, 1–2, 17–22
  as a critical technology,
    185–192
  as a technological paradigm,
    51–58, 106–108
  as a technological revolution,
    2–3
  as body of knowledge and
    skill, 52–53
  as interrelationships with
    industries, 55–58
  as systems of devices, 21–22,
    53–55
  compared to biotechnology, 33,
    39
  compared to electronics, 16–17
  consequences for industries, 2
  declining U.S. performance in,
    3–4
  effects on industrial sectors,
    34–35, 55–57, 247 n.69,
    253 n.22
  history of, 22–31
  industry-university joint
    research centers in,
    145–146
  large firms in, 31
  market value, 32–33, 34–35,
    246 n.69
  numbers of firms, 3, 31
  numbers of products, 246 n.63
  origins in nonmarket
    institutions, 30–31
  origins of the word, 16, 240
    n.3
  prognostications about, 1–2
  R&D investment in, 4
  technical introductions to, 239
    n.1
  U.S. trade performance, 36–39
Picking winners, 97–98, 100
Planning
  definition of, 254 n.30
  and common resources, 65
  flaws of, 237
  for industrial change, 8, 61
  in industry-university joint
    research, 172
  in the contract state, 85–86
  in U.S. military agencies,
    217–222
  institutional capability for, 236
  paradox of, 235
  U.S. inability in, 6

Pluralism, 101
Policy research, 233–234
Political economy of capitalism,
  81–88, 109
  and national industrial
    advantage, 234
  comparative studies of, 81–82
  depiction of, as free market, 9
Pollicove, Harvey, 204, 206
Polytechnic University, 145, 163,
  165
Pork barrel appropriations, 101,
  116–123, 138
  examples of, 122
  for optoelectronic materials
    center, 199
  for science buildings, 116–123,
    266 n.31
  "have and have-not"
    campuses, 121, 266 n.30
  in New York State, 119–121
  lobbying by academic
    institutions for, 119–123
Precision optics, 200–207, 218
U.S. industrial decline in, 202
U.S. trade performance in, 37
Price, Don, 84–86, 230
Princeton University, 145, 156,
  158
Privatization, 5–6, 9, 14, 79–81,
  84–86, 108–111, 227
  meanings of, 79–80
  disaggregative form, 138–139
  failures of, 6
  forms of, 11–13, 231–232
  industrial consortia, 144
  industry-university joint
    research, 173–174
  in policy making, 80, 230–233
  military programs, 225–226
  reform of, 236, 261 n.37, 289 n.8
Public goods, 67–71, 104
Public infrastructure, 88
Public policy, implications of this
  study for, 234–238

Quantum electronics, 30

R&D tax credit, 114–115, 263
  n.2
Reagan (presidential)
  administration, 78, 132
Regulatory policy, 259 n.37
Reich, Robert, 77, 78, 89–90
Reis, Victor, 216
Rensselaer Polytechnic Institute,
  119
Republican (party) politics, 232
Research agendas (in industry-
  university joint research),
  141–142, 163–174
  ambiguous dynamics of, 164
  and patenting, 166–167
  business participants in,
    164–169, 276 n.73
  faculty participants in,
    167–169
  pressures on, 164, 169–173
  strategic planning for, 169–173
Rochester (New York), 7, 57, 205,
  253 n.25
Rochester Institute of
  Technology, 117–119, 121,
  145, 165
Rome Air Development Center
  (Griffiss Air Force Base, New
  York), 120, 220
Rose, M. Richard, 118, 121
Rosenberg, Nathan, 92, 104–106
Russia. *See* Soviet Union
Rutgers University, 145,
  155–156, 164, 166

Sandia National Laboratories, 200
Savas, E.S., 79
Schawlow, Arthur, 24
Schultze, Charles L., 95–100
Schumpeter, Joseph, 49, 92, 252
  n.13
Science policy. *See* Technology
  policy

Scientific paradigms, 46–47, 251 nn.6, 7
Scope (of this book), 7–8
Scott, Bruce R., 75
SDI *See* Strategic Defense Initiative
Semiconductor lasers, 27–28, 36
Semiconductor Manufacturing Technology Consortium (Sematech), 180
Sensing devices, 20
Shannon, Robert, 163, 275 n.56
Sheltering (*see also* DARPA, Military industrial programs), 13, 175–228, 231–232
  as a political formula, 216–217, 225–226
  as form of privatized industrial policy, 225–226
Signature control, 187
Silber, John, 121
Skills, of master opticians, 203
Society for Imaging Science and Technology, 161
Sources of information (for this book), 6–7
Soviet Union, 28, 177
  in photonics, 191
SPIE. *See* International Society for Optical Engineering
Spinoffs, technological, 193–194, 196–197
Standards, technological, 124
Star Wars. *See* Strategic Defense Initiative
State capitalism, 82–83
State University of New York at Buffalo, 119
Strategic Defense Initiative, 192–197, 200
  civilian spinoffs, 193–194, 196–197
  programs in electro-optics, 194
  whether it was industrial policy, 200

Strategic materials, 199
Strategic Optoelectronics Materials Center (proposed), 198–200
Strategic industries
  as argument for industrial policy, 97
  knowledge of, in industrial policy making, 108
Strategic trade policy, 99
Strategy. *See* planning
Suh, Nam P., 132–133, 153
Sununu, John, 215
Switzerland, 82, 83
Systems, technical
  compared to integrality, 54–55
  in technological paradigms, 53–55

Taft IV, William, 204
Tassey, Gregory, 39, 126, 127
Tax Reform Act of 1986, 114
Technological paradigm, 4–5, 46–51
  definition of, 251 n.11, 252 n.20
  and nonmarket institutions, 61
  as a common resource, 8, 57–58, 63–64
  as body of knowledge and skill, 52–53, 60
  as cause of industrial dislocations, 56–57
  as interrelationships with industries, 55–58, 60–61, 253 n.23
  as justification for public intervention, 59–61
  as resource for industries, 53
  as technical system, 21–22, 53–55, 60
  characteristics of, 8, 48–49
  compared to technological discreteness, 5, 110
  diagraming, 53, 252 n.17

Technological paradigm
  continued
  electrification as, 54
  in national industrial
    advantage, 63
  in technological revolutions,
    49–50
  indicators of, 51–52, 254 n.27
  integrality of, 50–51
  photonics as example of, 4–5,
    51–58
  predictability of, 10, 105–108
Technological revolutions, 2–3
  comparative structures of, 234
  role of paradigms in, 49–50, 51
Technology policy (*See also*
    Industrial policy)
  and the misunderstanding of
    technology transfer, 61–63
  and types of capitalism, 83
  compared to industrial policy,
    91–94
  criteria for, 5
  post-war, 85–86
Technology transfer, 61–63, 115
  contrasted with concept of
    paradigm, 62–63
  military-industrial, 62
Technology Transfer Act of 1986,
  115, 129
Thurow, Lester, 89–90, 95–96,
  101
Townes, Charles, 24
Tyson, Laura, 90

U.S. Department of Commerce,
  38, 87–88, 91, 126, 203
U.S. Department of Defense (*See
  also* Critical technologies
  lists, DARPA, Military
  industrial programs)
  ability to mobilize, 175
  and precision optics, 201–207
  coordination of technical
    programs, 182–184,
    221–222

dependence on foreign
    sources, 175
  development of fiber optics, 28
  development of imaging,
    29–31
  development of lasers, 24–25
  optoelectronic materials
    initiative, 197–198
  research expenditure on
    optoelectronics, 39–40
U.S. Department of Energy, 115,
  119, 129
U.S. International Trade
  Commission, 36
U.S. Patent and Trademark
  Office, 37
United Kingdom, 38, 83
University of Alabama, 118–119,
  121, 145, 146, 196, 197
University of Arizona, 145, 150,
  162, 166, 275 n.56
University of California, Santa
  Barbara, 145, 150
University of Central Florida,
  145, 157
University of Colorado, 145, 151,
  163, 168
University of Illinois at Urbana-
  Champaign, 145, 151,
  170–171, 172
University of Michigan, 145, 150
University of New Mexico,
  199–200
University of Rochester,
  119–120, 145, 146, 159–162,
  165, 167, 168
  Center for Advanced Optical
    Technology, 157, 159–162,
    168
  Center for Optics
    Manufacturing, 162, 201,
    204–207
  Institute of Optics, 23, 159, 162
University-industry joint
  research. *See* Industry-
  university joint research

Urban land policy, 81
USS *Vincennes*, 212

Very high speed integrated
  circuit development
  program, 179
Virginia Center for Innovative
  Technology, 155
Virginia Polytechnic Institute,
  146, 155

West Germany, 37–38, 75, 83,
  204

Whinnery, John R., 36
White, Robert, 151–152
Wildavsky, Aaron, 89, 101

Xerox Corp., 32, 57, 143, 157,
  160, 271 n.12

Yang, Andrew C., 209

Zenith Electronics Corp., 213,
  214
Zysman, John, 69–70, 82–84,
  86, 90, 97